MASTERING THE GEORGIA 5th GRADE CRCT

IN

MATHEMATICS

Developed to the new Georgia Performance Standards!

ERICA DAY
COLLEEN PINTOZZI
TANYA KELLEY

AMERICAN BOOK COMPANY
P. O. BOX 2638
WOODSTOCK, GEORGIA 30188-1383
TOLL FREE 1 (888) 264-5877 PHONE (770) 928-2834
FAX (770) 928-7483
WEB SITE: www.americanbookcompany.com

Acknowledgements

In preparing this book, we would like to acknowledge Mary Stoddard for her contributions in editing and developing graphics, and John Martino and Philip Jones for their contributions in writing for this book. We would also like to thank our many students whose needs and questions inspired us to write this text.

by American Book Company
P.O. Box 2638
Woodstock, GA 30188-1383

Printed in the United States of America
11/06

Contents

Preface

Passing the Georgia 5th Grade CRCT in Mathematics will help you review and learn important concepts and skills related to elementary school mathematics. First, take the Diagnostic Test beginning on page 1 of the book. To help identify which areas are of greater challenge for you, complete the evaluation chart with your instructor in order to help you identify the chapters which require your careful attention. When you have finished your review of all of the material your teacher assigns, take the progress tests to evaluate your understanding of the material presented in this book. **The materials in this book are based on the Georgia Performance Standards including the content descriptions for mathematics, which are published by the Georgia Department of Education. The complete list of standards is located in the Answer Key. Each question in the Diagnostic and Practice Tests is referenced to the standard, as is the beginning of each chapter.**

This book contains several sections. These sections are as follows: 1) A Diagnostic Test; 2) Chapters that teach the concepts and skills for ***Passing the Georgia 5th Grade CRCT in Mathematics***; and 3) Two Practice Tests. Answers to the tests and exercises are in a separate manual.

ABOUT THE AUTHORS

Erica Day has a Bachelor of Science Degree in Mathematics and is working on a Master of Science Degree in Mathematics. She graduated with high honors from Kennesaw State University in Kennesaw, Georgia. She has also tutored all levels of mathematics, ranging from high school algebra and geometry to university-level statistics, calculus, and linear algebra. She is currently writing and editing mathematics books for American Book Company, where she has coauthored numerous books, such as ***Passing the Georgia Algebra I End of Course***, ***Passing the Georgia High School Graduation Test in Mathematics***, and ***Passing the New Jersey HSPA in Mathematics***, to help students pass graduation and end of course/grade exams.

Colleen Pintozzi has taught mathematics at the middle school, junior high, senior high, and adult level for 22 years. She holds a B.S. degree from Wright State University in Dayton, Ohio and has done graduate work at Wright State University, Duke University, and the University of North Carolina at Chapel Hill. She is the author of many mathematics books including such best-sellers as *Basics Made Easy: Mathematics Review, Passing the New Alabama Graduation Exam in Mathematics, Passing the Louisiana LEAP 21 GEE, Passing the Indiana ISTEP+ GQE in Mathematics, Passing the Minnesota Basic Standards Test in Mathematics,* and *Passing the Nevada High School Proficiency Exam in Mathematics.*

Tanya Kelley graduated cum laude from Lehman College in New York with a Bachelor of Arts in Accounting. She has developed a money management program, which emphasizes fundamental math skills and serves both children and adults. She is currently the Finance Manager for American Book Company and is a member of the National Honors Society.

Formula Sheet

Circumference	Circle	$C = 2\pi r$ or $C = \pi d$
Area	Rectangle	$A = lw$ or $A = bh$
	Parallelogram	$A = bh$
	Triangle	$A = \frac{1}{2} bh$ or $A = \frac{bh}{2}$
	Circle	$A = \pi r^2$
Volume	Cube	$V = 6s$
	Rectangular Prism	$V = Bh$*
	** B represents the area of the Base of a solid figure.*	
Pi	π	$\pi \approx 3.14$

Converting Units of Measure		**Abbreviations**
Volume	1 gallon = 4 quarts	gallon = gal
	1 quart = 2 pints	quart = qt
	1 pint = 2 cups	pint = pt
	1 cup = 8 ounces	ounce = oz
Length	1 mile = 5,280 feet	mile = mi
	1 yard = 3 feet	yard = yd
	1 foot = 12 inches	foot = ft
		inches = in
Weight	16 ounces = 1 pound	pound = lb

Diagnostic Test

1. Which of the following is a composite number?

   A. 4
   B. 13
   C. 2
   D. 7

   M5N1a

2. Which of the following is a prime number?

   A. 7
   B. 21
   C. 4
   D. 18

   M5N1a

3. What fraction does the shaded area represent?

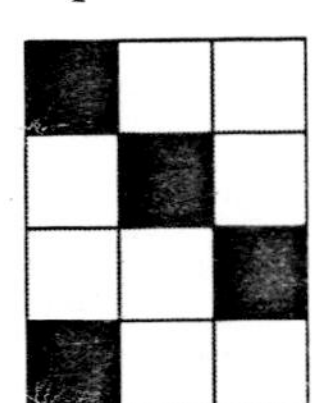

   A. $\frac{4}{12}$
   B. $\frac{2}{3}$
   C. $\frac{1}{2}$
   D. $\frac{5}{12}$

   M5N4e

4. Which number is a multiple of 6?

   A. 3
   B. 24
   C. 27
   D. 2

   M5N1b

5. Which number is a factor of 39?

   A. 13
   B. 6
   C. 78
   D. 19

   M5N1b

6. The shaded area represents what percentage?

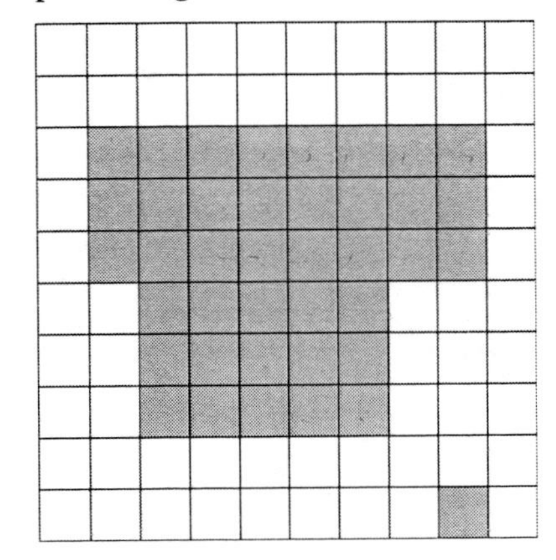

   A. 45%
   B. 40%
   C. 42%
   D. 4%

   M5N5a

7. Which number is divisible by 2, 4, and 5?

   A. 10
   B. 30
   C. 40
   D. 12

   M5N1c

8. The numbers 3, 6, and 18 are all divisible by which number?

   A. 2
   B. 9
   C. 3
   D. 6

   M5N1c

9. What is the place value of the underlined digit?

5,<u>4</u>16,032

A. forty-thousand
B. four million
C. four thousand
D. four hundred thousand

M5N2a

10. The unshaded area represents what percentage?

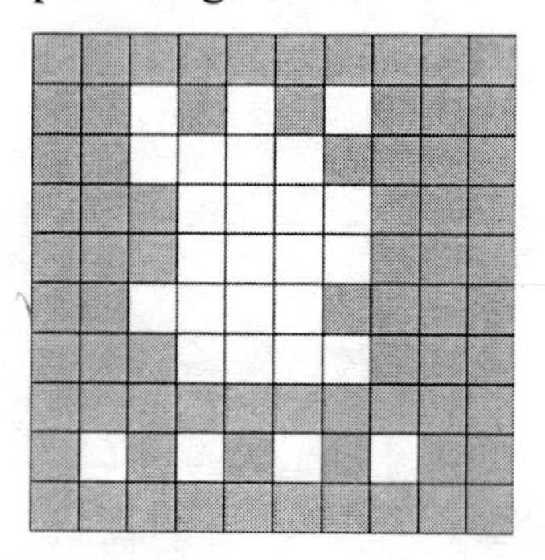

A. 73%
B. 27%
C. 37%
D. 70%

M5N5a

11. What is the place value of the underlined digit?

84,167.0<u>4</u>8

A. four hundred
B. four thousandths
C. four hundredths
D. four tenths

M5N2a

12. Find the product.

$5,482 \times 100 =$

A. 54,820
B. 5,482,000
C. 504,820
D. 548,200

M5N2b

13. What fraction does the shaded portion represent?

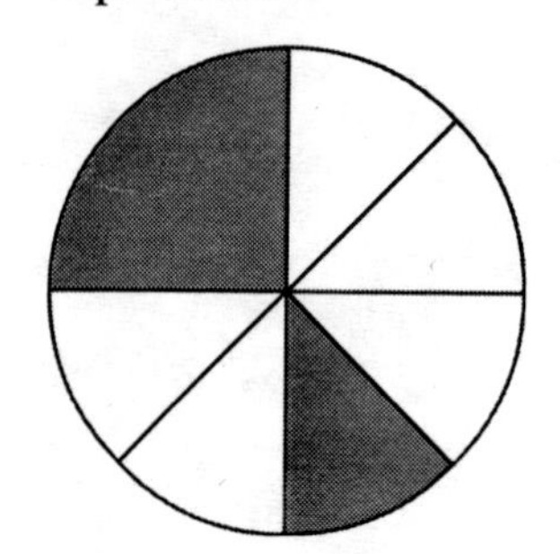

A. $\frac{1}{4}$
B. $\frac{3}{8}$
C. $\frac{4}{8}$
D. $\frac{2}{4}$

M5N4e

14. Find the product.

$8,714 \times 0.01 =$

A. 871.40
B. 8.714
C. 87,140
D. 87.14

M5N2b

15. Find the product.

479.1×0.21

A. 100.51
B. 1,006.1
C. 1,061.0
D. 100.611

M5N3a

16. Estimate the area.

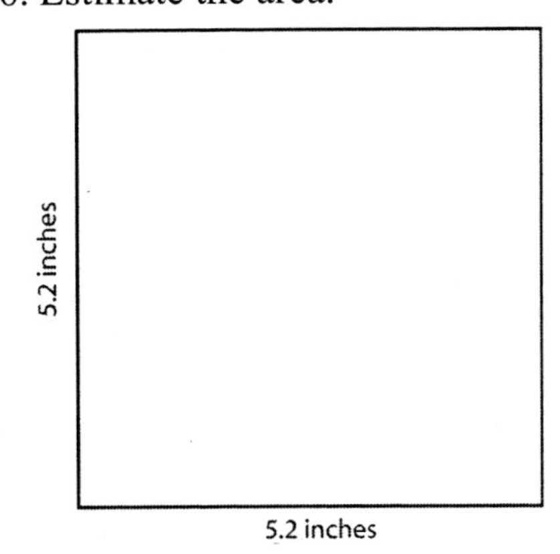

A. 20 in^2
B. 25 in^2
C. 10 in^2
D. 15 in^2

M5M1a

17. Find the area.

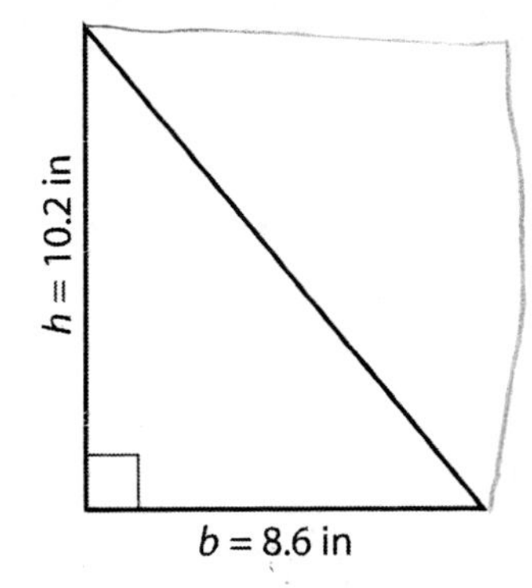

A. 87.72 in^2
B. 94.72 in^2
C. 43.86 in^2
D. 27.40 in^2

M5M1d

18. Find the product.

287×0.14

A. 4,018
B. 401.8
C. 40.18
D. 408

M5N3c

19. Find the product.

$1,465 \times 2.6$

A. 3,809
B. 3,709
C. 380.9
D. 370.9

M5N3c

20. $871.4 \div 2.1 =$ _____

A. 405
B. 411.95
C. 414.95
D. 404.95

M5N3d

21. Find the area of the irregular polygon shown in the diagram below.

A. 11 sq. units
B. 22 sq. units
C. 28 sq. units
D. 30 sq. units

M5M1f

22. $651 \div 0.4 =$ _____

A. 1627.5
B. 1626.5
C. 260.40
D. 1027.5

M5N3d

23. $2,128 \div 4 =$ _____

A. 532
B. 57
C. 507
D. 517

M5N3d

24. Find the area.

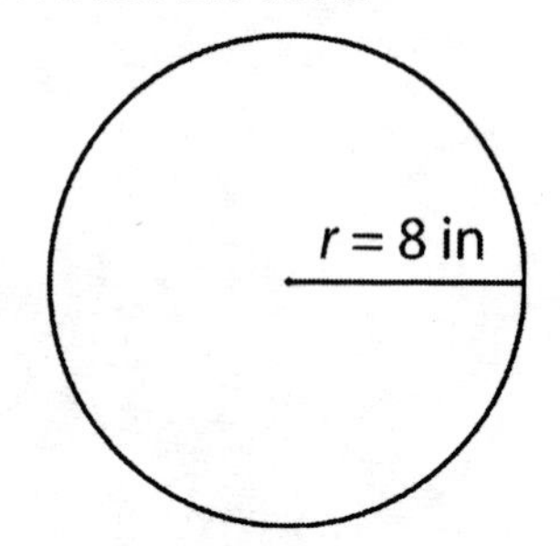

A. 200.96 in^2
B. 25.12 in^2
C. 50.24 in^2
D. 12.06 in^2

M5M1e

25. Find the quotient: $165 \div 5$

A. 31
B. 33
C. 21
D. 23

M5N4a

26. Simplify the fraction: $\frac{6}{8}$

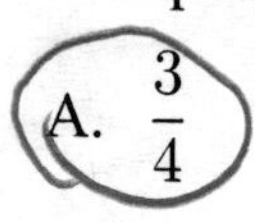

A. $\frac{3}{4}$
B. $\frac{1}{2}$
C. $\frac{4}{5}$
D. $\frac{1}{4}$

M5N4c

27. Find the area.

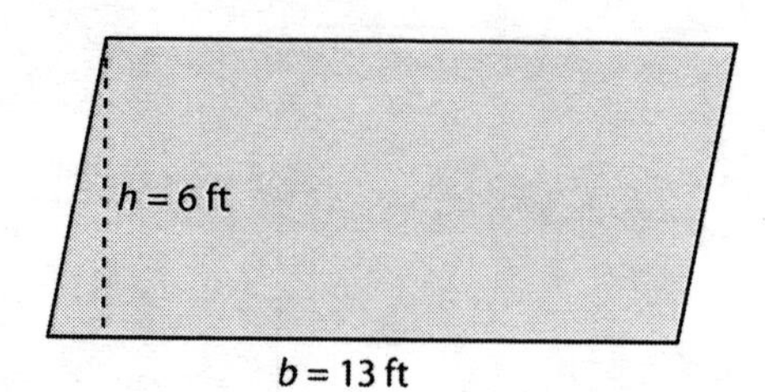

A. 19 ft^2
B. 78 ft^2
C. 78 ft
D. 19 ft

M5M1d

28. Find the product: $\frac{4}{5} \times \frac{4}{4} =$

A. 4
B. $\frac{4}{5}$
C. $\frac{8}{9}$
D. $1\frac{1}{4}$

M5N4b

29. _____ $> \frac{3}{4}$

A. $\frac{4}{5}$
B. $\frac{3}{5}$
C. $\frac{4}{8}$
D. $\frac{5}{10}$

M5N4f

30. Find the area of the polygon shown in the diagram below.

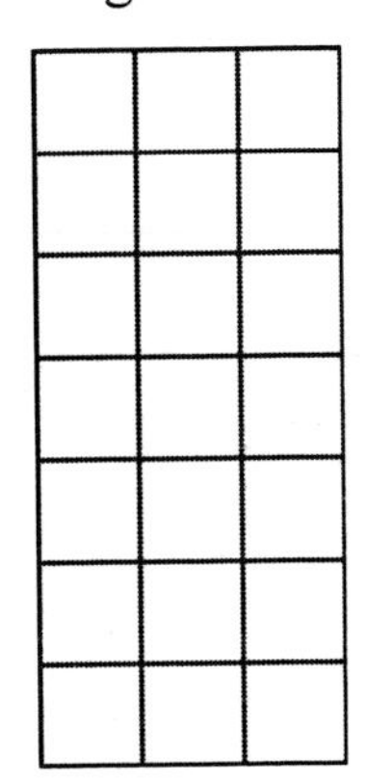

A. 10 sq. units
B. 20 sq. units
C. 21 sq. units
D. 49 sq. units

M5M1f

31. $\frac{6}{12}$ is equivalent to which fraction?

A. $\frac{4}{6}$

B. $\frac{18}{36}$

C. $\frac{2}{24}$

D. $\frac{1}{4}$

M5N4c

32. Solve. $\frac{12}{16} - \frac{3}{8} =$

A. $\frac{1}{2}$

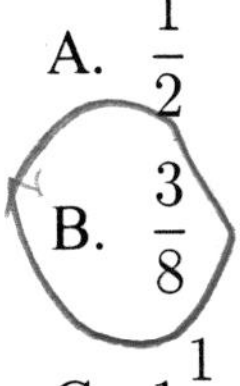

B. $\frac{3}{8}$

C. $1\frac{1}{8}$

D. $1\frac{1}{4}$

M5N4c

33. $\frac{4}{8} + \frac{6}{10} =$ ___

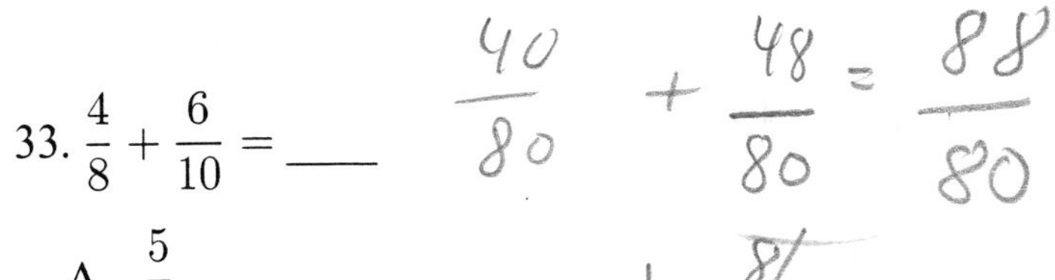

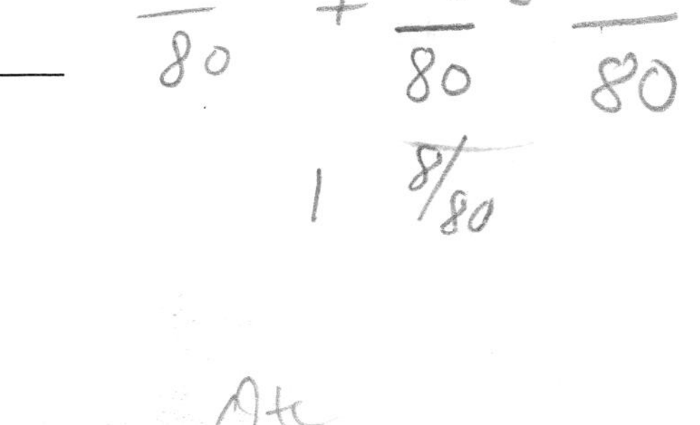

A. $\frac{5}{9}$

B. $1\frac{1}{10}$

C. $\frac{4}{7}$

D. $\frac{11}{20}$

M5N4g

34. Find the area.

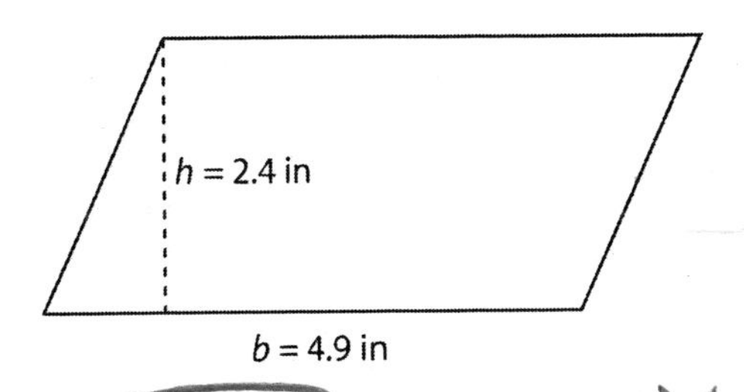

A. 10.76 in^2
B. 11.96 in^2
C. 14.6 in^2
D. 11.76 in^2

M5M1d

35. $6\frac{1}{7} - 5\frac{2}{3} =$___

A. $\frac{10}{21}$

B. $1\frac{1}{3}$

C. $\frac{11}{21}$

D. $1\frac{1}{7}$

M5N4g

36. Estimate: 47×21

A. 987
B. 1,000
C. 800
D. 900

M5N4i

37. Estimate: $560 \div 12$

A. 5,000
B. 600
C. 60
D. 47

M5N4i

38. $\frac{4}{6}$ is not equivalent to ______.

A. 461
B. 0.66
C. $\frac{2}{3}$
D. $\frac{66}{100}$

M5N4c

39. $\frac{6}{10}$ = ______.

A. 0.06
B. 0.60
C. 6
D. 0.006

M5N4h

40. $\frac{7}{5}$ = ______.

A. 1.25
B. 1.40
C. 1.05
D. 1.50

M5N4h

41. Estimate: $5,216 \times 21$

A. 100,000
B. 120,000
C. 109,536
D. 10,000

M5N4i

42. How was Figure A moved?

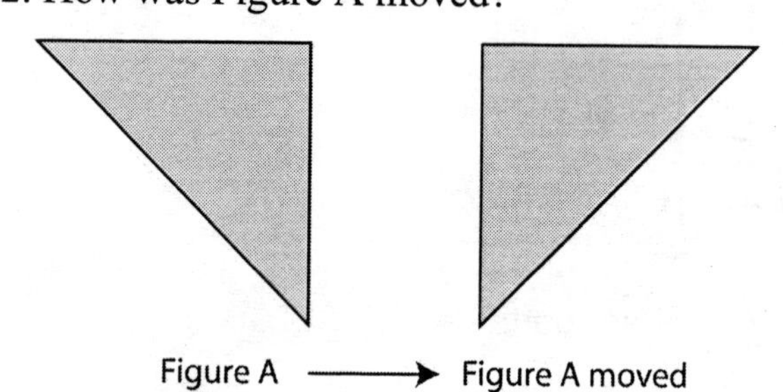

A. Flipped
B. Turned
C. Slid & Turned
D. Slid only

M5G1

43. Estimate: $151,217 \div 32,114$

A. 3
B. 4.71
C. 50
D. 5

M5N4i

44. Which measurement is the largest?

A. 3 L
B. 30 mL
C. 0.3 mL
D. 300 mL

M5M3b

45. Find the volume of a rectangular prism, measured as follows:

$l = 12.8$ yds, $w = 4$ yds, $h = 9.2$ yds

A. 26 yd^3
B. 47.1 yd^3
C. 471.04 yd^3
D. 51.2 yd^3

M5M4d

46. $2\frac{3}{8}$ = _____

A. $\frac{19}{8}$

B. $\frac{30}{8}$

C. $\frac{6}{8}$

D. $\frac{3}{16}$

M5N4h

47. $\frac{2}{5} \times \frac{1}{2}$ = _____

A. $\frac{2}{10}$

B. $\frac{40}{20}$

C. $\frac{2}{5}$

D. $\frac{1}{4}$

M5N4d

48. $\frac{5}{8}$ divided by $\frac{1}{10}$ =

A. $\frac{1}{16}$

B. $6\frac{1}{4}$

C. $5\frac{1}{4}$

D. $\frac{5}{80}$

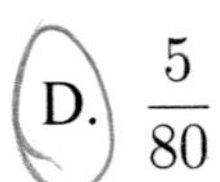

M5N4d

Use the graph to answer the questions 49 and 50.

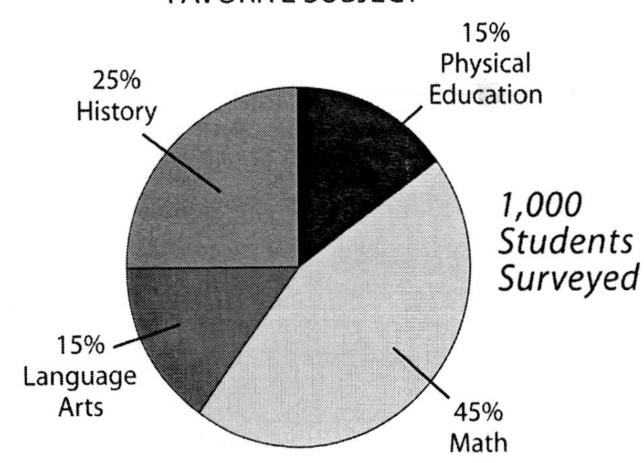

49. How many more students prefer math to history?

A. 450
B. 250
C. 200
D. 300

M5N5b

50. How many students prefer physical education and language arts?

A. 150
B. 300
C. 200
D. 350

M5N5b

51. 18 L = _____ mL

A. 18,000
B. 1,800
C. 180,000
D. 180

M5M3a

52. 17 gallons = _____ quarts

A. 8
B. 34
C. 68
D. 4.25

M5M3b

53. Find the volume.

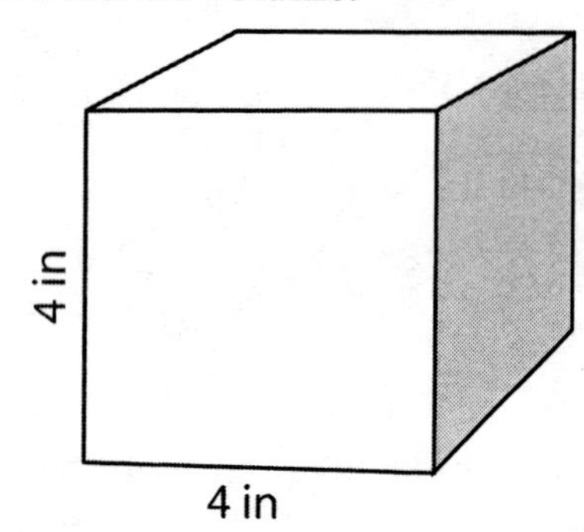

A. 16 in^3
B. 64 in^3
C. 64 in^2
D. 16 in

M5M4d

54. Find the volume.

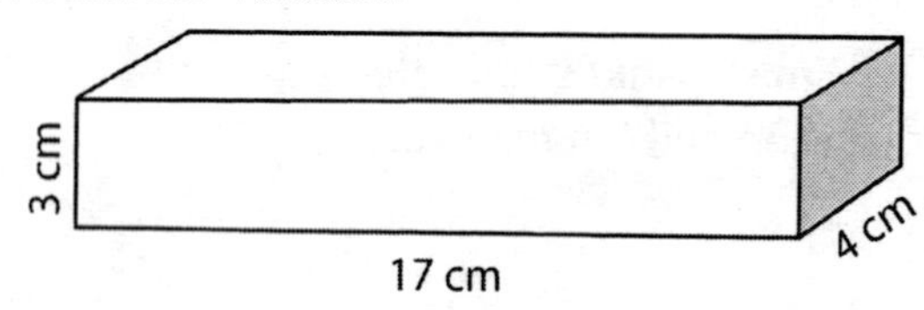

A. 68 cm^2
B. 68 cm
C. 204 cm^3
D. 24 cm^3

M5M4d

55. Estimate the volume of a rectangular prism, with measurements as follows:

$l = 5.1$ cm, $w = 6.2$ cm, $h = 8.4$cm

A. 20 cm^3
B. 240 cm^3
C. 265.61 cm^3
D. 197 cm^3

M5M4e

56. What is the capacity of a container that has a length of 6 cm, width of 4 cm, and a height of 3 cm?

A. 72 mL
B. 24 mL
C. 36 mL
D. 13 cm

M5M4f

57. If $\angle ABC$ is congruent to $\angle DEF$, what is the measurement of $\angle DEF$?

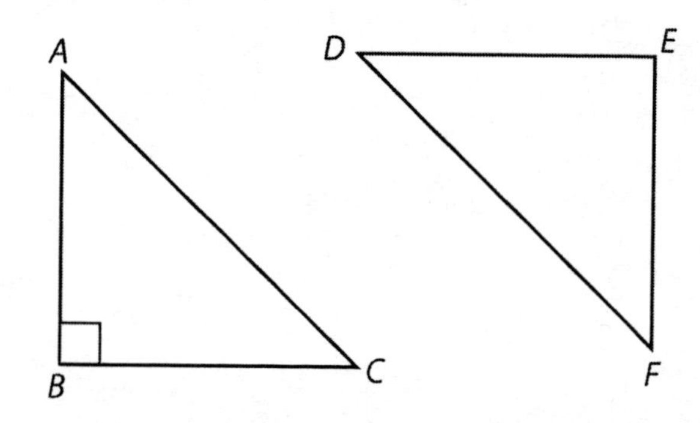

A. 100 degrees
B. 90 degrees
C. 95 degrees
D. none of the above

M5G1

58. Find the circumference of a circle with a diameter of 8 inches.

A. 40.24 in
B. 24.12 in
C. 25.12 in
D. 50.24 in

M5G2

59. Find the circumference of a circle with a radius of 4.6 inches.

A. 66.44 in
B. 14.44 in
C. 28.89 in
D. 12.56 in

M5G2

60. What is the value of the expression $5x + 6$ when $x = 3$?

A. 9
B. 15
C. 21
D. 45

M5A1b

Use the graph below to answer questions 61 and 62.

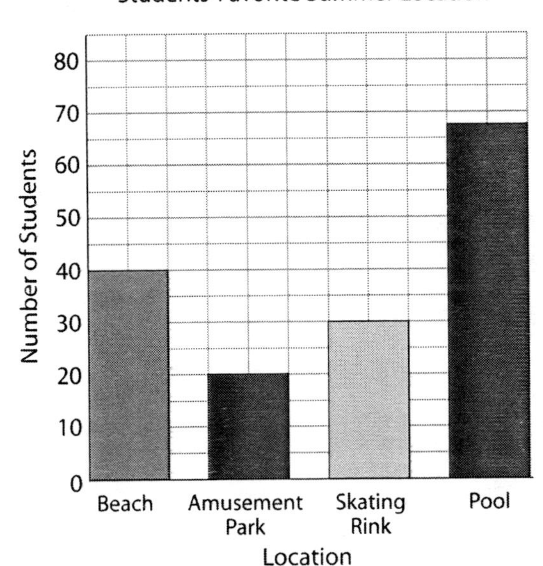

61. About how many students were surveyed?

A. 150
B. 157
C. 167
D. 170

M5D1a

62. If the beach selection is expected to increase by 50% next year, how many students will that include?

A. 50
B. 60
C. 20
D. 80

M5D1a

63. Translate the following sentence into an algebraic equation:

The sum of a number, x, and 6 is 14.

A. $x + 6 = 14$
B. $x - 6 = 14$
C. $x \times 6 = 14$
D. $x \div 6 = 14$

M5A1a

64. Sandy would like to know if hot lunch sales have increased over the last three years. Which graph would **not** be used to display the outcome?

A. line
B. circle
C. bar
D. all of the above

M5D1b

65. Which color was the most popular?

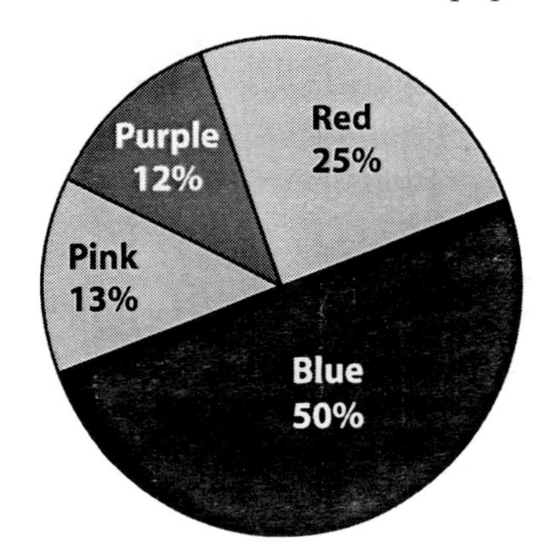

A. red
B. blue
C. purple
D. pink

M5D1a

66. Which sale item offers the best buy per piece?

A. bananas
B. apples
C. plums
D. oranges

M5P1a

67. Azrina bought a rectangular fish tank with the following dimensions.

$l = 18$ in, $w = 10$ in, $h = 12$ in

What is the volume?

A. 2,160 in^2
B. 180 in^2
C. 180 in^3
D. 2,160 in^3

M5P1b

68. Rob & Ryan went to the skating rink. They each bought 2 sodas, 1 skate rental, and 2 hot dogs. If they each had $6.00 remaining, how much money did they initially have combined?

Soda	$1.50
2 Hot Dogs	$4.00
Skate Rental	$5.00
Candy	$3.00
Hamburgers	$4.00

A. $12.00
B. $36.00
C. $18.00
D. $20.00

M5P1b

Use the graph below to answer questions 69 and 70.

My Time
Movie Theatre

All movies before 6 pm $5.00

6 pm - closing:

Adults	$7.50
Children	6.00

69. The local movie theater earned a total of $1,455 in ticket sales this weekend. If 52 adults and 10 children attended after 6 p.m., how many people took advantage of the bargain matinee?

A. 104
B. 52
C. 201
D. 104

M5P1c

70. If the theater predicts a 20% reduction in sales next weekend, what will anticipated sales equal?

A. $1,164
B. $964
C. $991
D. $1,114

M5P1a

Evaluation Chart for the Diagnostic Mathematics Test

Directions: On the following chart, circle the question numbers that you answered incorrectly. Then turn to the appropriate topics, read the explanations, and complete the exercises. Review the other chapters as needed. Finally, complete the ***Passing the Georgia 5th Grade CRCT in Mathematics*** Practice Tests to further review.

		Questions	Pages
Chapter 1:	Whole Numbers and Number Sense	1, 2, 4, 5, 7, 8, 9, 12, 23, 25	13–38
Chapter 2:	Decimals	11, 14, 15, 18, 19, 20, 22	39–52
Chapter 3:	Fractions	26, 28, 29, 31, 32, 33, 35, 36, 37, 38, 39, 40, 41, 43, 46, 47, 48	53–72
Chapter 4:	Percents	3, 6, 10, 13	73–85
Chapter 5:	Introduction to Algebra	60, 63	86–92
Chapter 6:	Measurement	44, 51, 52	93–101
Chapter 7:	Plane Geometry	16, 17, 21, 24, 27, 30, 34, 42, 57, 58, 59	102–119
Chapter 8:	Solid Geometry	45, 53, 54, 55, 56, 67	120–126
Chapter 9:	Data Interpretation	49, 50, 61, 62, 64, 65, 66, 68, 69, 70	127–139

NOTES

Chapter 1
Whole Numbers and Number Sense

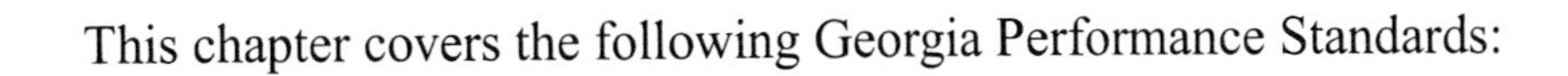
This chapter covers the following Georgia Performance Standards:

M5N	Numbers and Operations	M5N1.a, b, c M5N2.a
M5P	Process Skills	M5P1.a, b, c M5P2.a, b

1.1 Place Value: Greater Than One

Place Value: The value of a digit based upon its place, within the number. For example, in the chart below, the number 5 has two very different values in the number 987,654,325:

Hundred-million 100,000,000	Ten-million 10,000,000	Million 1,000,000	Hundred-thousand 100,000	Ten-thousand 10,000	Thousand 1,000	Hundred 100	Tens 10	Ones 1.0
9	8	7	6	5	4	3	2	5

From the right, let's start in the ones column. As we progress from right to left, we see that each column represents 10 times the value of the column before it:

Ones:	1
Tens:	10 times 1 =10
Hundreds:	10 times 10 = 100
Thousands:	10 times 100 = 1000
Ten-Thousands:	10 times 1000 = 10,000
Hundred-Thousand:	10 times 10,000 = 100,000
Million:	10 times 100,000 = 1,000,000
Ten Million:	10 times 1,000,000 = 10,000,000
Hundred-Million:	10 times 10,000,000 = 100,000,000

Using the place value chart on the previous page:

From right to left, the first five, listed in the ones column represents 5 ones.

$5 \text{ times } 1 = 5$

Continuing to the second five from the right, we see that that five has a very different value:

$5 \text{ times } 10,000 = 50,000$.

Let's write this number in expanded form to clearly identify the value of each number. (Again, notice the different values of the number five.)

The number written in **standard** form: $987,654,325$

The number written in **expanded** form: $900,000,000 + 80,000,000 + 7,000,000 + 600,000 + 50,000 + 4,000 + 300 + 20 + 5 = 987,654,325$

We read it as follows: nine hundred eighty-seven million, six hundred fifty- four thousand, three hundred twenty- five.

Write the value of the number 4. The first one has been done for you.

1. 14,167   Four thousand (4,000)
2. 304,625
3. 64
4. 45,678,123
5. 46
6. 114,868,012
7. 427,658,923
8. 4,167

Use the number below to answer the following questions.

847,652,109

Which number is in the

9. Tens place?
10. Millions place?
11. Hundred-thousands place?
12. Hundred-millions place?
13. Hundreds place?
14. Thousands place?
15. Ten-thousands place?
16. Ten-millions place?

Write the value of the underlined digit. The first one has been done for you.

17. 21<u>5</u>   five ones
18. <u>3</u>,257,866
19. 6<u>6</u>0
20. <u>2</u>0,547
21. <u>6</u>18,798,622
22. <u>1</u>42,661
23. <u>8</u>9
24. 1,6<u>2</u>7

1.2 Even and Odd Numbers

Even Numbers are numbers that can be divided into two groups, without a remainder and end in the digits 0, 2, 4, 6, or 8.

Example 1: The number 10 is an even number. 10 is even because it can be broken up into 5 and 5 ($5 + 5 = 10$), which are two groups with the same number in each group.

Example 2: Is 22 an even number? Why or Why not?

Yes, 22 is an even number. Here's why...

The number ends in 2.

Therefore, 22 is an even number.

Odd Numbers are numbers that can not be divided into two groups without a remainder and end in the digits 1, 3, 5, 7, or 9.

Example 3: The number 11 is an odd number. 11 cannot be broken up into two groups evenly. The closest you can get is 5 and 6 ($5 + 6 = 11$).

Example 4: Is 9 an odd number? Why or why not?

Yes, 9 is an odd number. Here's why...

The number ends in 9.

Therefore, 9 is an odd number.

Circle the even numbers and put a line under the odd numbers.

1. 16
2. 2
3. 9
4. 27
5. 15
6. 7
7. 11
8. 108
9. 315
10. 100
11. 5
12. 13

1.3 Practicing Adding

When adding numbers, you must carry whenever a column is over ten. For example, if you add 9 and 8, the answer is 17. You will write the seven down and carry the one to the next column. Then you will add it to the numbers in that column.

Example 5: Add 7,429 and 3,618.

Step 1: Remember when you add to arrange the numbers in columns with the ones digits at the right.

$$\begin{array}{r} 7429 \\ +\ 3618 \\ \hline \end{array}$$

Step 2: Start at the right and add each column. Remember to carry when necessary.

$$\begin{array}{r} {}^{1\ \ 1} \\ 7,429 \\ +\ \ 3,618 \\ \hline 11,047 \end{array}$$

Find the sum, and circle your answer.

1. $\begin{array}{r} 6782 \\ +\ 4009 \\ \hline \end{array}$

2. $\begin{array}{r} 6009 \\ +\ 5872 \\ \hline \end{array}$

3. $\begin{array}{r} 9864 \\ +\ 7643 \\ \hline \end{array}$

4. $\begin{array}{r} 3015 \\ +\ 496 \\ \hline \end{array}$

5. $\begin{array}{r} 3865 \\ +\ 984 \\ \hline \end{array}$

6. $\begin{array}{r} 3017 \\ +\ 486 \\ \hline \end{array}$

7. $\begin{array}{r} 8360 \\ +\ 4385 \\ \hline \end{array}$

8. $\begin{array}{r} 508 \\ +\ 9675 \\ \hline \end{array}$

9. $\begin{array}{r} 8400 \\ +\ 2509 \\ \hline \end{array}$

10. $\begin{array}{r} 6709 \\ +\ 891 \\ \hline \end{array}$

11. $\begin{array}{r} 1395 \\ +\ 8041 \\ \hline \end{array}$

12. $\begin{array}{r} 1467 \\ +\ 903 \\ \hline \end{array}$

13. $\begin{array}{r} 1921 \\ +\ 1492 \\ \hline \end{array}$

14. $\begin{array}{r} 8475 \\ +\ 3074 \\ \hline \end{array}$

15. $\begin{array}{r} 5049 \\ +\ 687 \\ \hline \end{array}$

1.4 Adding Whole Numbers

Example 6: Find $302 + 54 + 712 + 9$.

Step 1: Remember when you add to arrange the numbers in columns with the ones digits at the right.

$$\begin{array}{r} 302 \\ 54 \\ 712 \\ + \quad 9 \\ \hline \end{array}$$

Step 2: Start at the right and add each column. Remember to carry when necessary.

$$\begin{array}{r} {\scriptstyle 1} \\ 302 \\ 54 \\ 712 \\ + \quad 9 \\ \hline 1,077 \end{array}$$

Find the sum, and circle your answer.

1. $18 + 24 + 157$
2. $2,458 + 5,011$
3. $4,005 + 1,342$
4. $386 + 54 + 3$
5. $4,057 + 21 + 219$
6. $2,465 + 486$
7. The total of 9 and 104
8. 94 more than 541
9. 784 increased by 51
10. 18 more than 149
11. 5 more than 557
12. 102 added to 73
13. 298 increased by 25
14. 541 plus 402
15. $12 + 454 + 3 + 97$
16. The sum of 308 and 52
17. The total of 85, 78, and 215
18. $6 + 243 + 19$

1.5 Practicing Subtraction

When subtracting numbers, you must borrow whenever the top number is smaller than the bottom number.

Example 7: Find 5001 − 982

Step 1: Remember when you subtract to arrange the numbers in columns with the ones digits at the right.

$$\begin{array}{r} 5001 \\ -982 \\ \hline \end{array}$$

Step 2: Start at the right, and subtract each column. Remember to borrow when necessary.

$$\begin{array}{r} 4\,9\,9 \\ \not{5}\not{0}\not{0}{}^{1}1 \\ -\;9\,8\,2 \\ \hline 4\,0\,1\,9 \end{array}$$

← Borrow 1 from the 500, making it 499.

Subtract. Questions 1, 6, and 11 have been done for you.

1. $\begin{array}{r} 5\,9\,9 \\ \not{6}\not{0}\not{0}{}^{1}5 \\ -\;9\,8\,9 \\ \hline 5\,0\,1\,6 \end{array}$

2. $\begin{array}{r} 8003 \\ -1815 \\ \hline \end{array}$

3. $\begin{array}{r} 5004 \\ -3747 \\ \hline \end{array}$

4. $\begin{array}{r} 9000 \\ -545 \\ \hline \end{array}$

5. $\begin{array}{r} 2007 \\ -339 \\ \hline \end{array}$

6. $\begin{array}{r} 2\,9 \\ \not{3}\not{0}{}^{1}0\,3 \\ -\;8\,9\,1 \\ \hline 2\,1\,1\,2 \end{array}$

7. $\begin{array}{r} 4006 \\ -762 \\ \hline \end{array}$

8. $\begin{array}{r} 8008 \\ -492 \\ \hline \end{array}$

9. $\begin{array}{r} 3007 \\ -144 \\ \hline \end{array}$

10. $\begin{array}{r} 7007 \\ -144 \\ \hline \end{array}$

11. $\begin{array}{r} 8\,9 \\ 1\,\not{9}\not{0}{}^{1}0\,2 \\ -\,4\,0\,1\,0 \\ \hline 1\,4\,9\,9\,2 \end{array}$

12. $\begin{array}{r} 20,000 \\ -19,814 \\ \hline \end{array}$

13. $\begin{array}{r} 50,001 \\ -4,692 \\ \hline \end{array}$

14. $\begin{array}{r} 2007 \\ -1872 \\ \hline \end{array}$

15. $\begin{array}{r} 8001 \\ -1496 \\ \hline \end{array}$

1.6 Subtracting Whole Numbers

Example 8: Find $1,006 - 568$

Step 1: Remember when you subtract to arrange the numbers in columns with the ones digits at the right.

$$\begin{array}{r} 1006 \\ -568 \\ \hline \end{array}$$

Step 2: Start at the right, and subtract each column. Remember to borrow when necessary.

$$\begin{array}{r} 9\,9 \\ \not{1}\not{0}\not{0}{}^{1}6 \\ -\underline{5\,6\,8} \\ 4\,3\,8 \end{array}$$ ← Borrow 1 from the 100, making it 99.

Note: When you see "less than" in a problem, the second number becomes the top number when you set up the problem.

Find the difference, and circle your answer.

1. $541 - 35$
2. $6007 - 279$
3. $694 - 287$
4. $902 - 471$
5. $500 - 376$
6. $1047 - 483$
7. 14 less than 607
8. 881 decreased by 354
9. The difference between 384 and 29
10. 560 decreased by 125
11. 43 less than 752
12. 74 less than 1,093
13. 96 less than 704
14. 327 less than 1,002
15. The difference between 273 and 55
16. The difference between 2,849 and 756
17. 975 decreased by 249
18. 405 decreased by 36

1.7 Multiplying Whole Numbers

Example 9: Multiply: 256×73

Step 1: Line up the ones digits. Multiply 256×3.

```
  1 1
  2 5 6
×   7[3]
-------
  7 6 8
```

- $3 \times 6 = 18$, write 8 and carry the one
- $3 \times 5 = 15$, add the 1 that was carried to get 16, write 6 and carry the one
- $3 \times 2 = 6$, add the 1 that was carried to get 7, write 7

Step 2: Multiply 256×7. Remember to shift the product one place to the left. Then add.

```
    3 4
    2 5 6
×   [7]3
---------
    7 6 8
  1 7 9 2
---------
1 8,6 8 8   ← Add
```

- $7 \times 6 = 42$, write 2 and carry the 4
- $7 \times 5 = 35$, add the 4 that was carried to get 39, write 9 and carry the 3
- $7 \times 2 = 14$, add the 3 that was carried to get 17, write 17

The answer is 18,688.

Multiply.

1. 258×72
2. 742×44
3. 785×32
4. 679×36
5. 841×27
6. 324×19
7. 921×23
8. 454×56
9. 156×95
10. 765×94
11. 581×25
12. 827×56
13. 942×24
14. 247×84
15. 468×43
16. 456×47
17. 743×65
18. 527×38
19. 524×39
20. 682×64

1.8 Divisibility Rules

Divisibility - A number that can be divided evenly, without a remainder.

Example 10: Is 12 divisible by 6?
Divide 6 into 12 to determine if the quotient has a remainder.
$12 \div 6 = 2 \text{ r } 0$
No remainder = Divisibility
This quotient does not have a remainder. Therefore, 12 is divisible by 6.

Example 11: Is 7 divisible by 3?
Divide 3 into 7 to determine if the quotient has a remainder.
$7 \div 3 = 2 \text{ r } 1$
This quotient does have a remainder.
Therefore, 7 is not divisible by 3.

Use the chart below to help you quickly determine if a particular number is divisible by 2, 3, 5, and/or 10.

Divisibility Rules Chart

A number is divisible by ...	if...
2	The last digit (ones place) is zero or an even number.
3	The sum of all the digits is divisible by 3. **Example:** $657 = 6 + 5 + 7 = 18$; $18 \div 3 = 6$. Therefore, 657 is divisible by 3.
5	The last digit (ones place) is a 0 or 5. **Example:** $1,125 \div 5 = 225$
10	The last digit is zero. (Example $20 \div 10 = 2$)

Review the numbers below. Write whether they are divisible by 2, 3, 5, or 10. You may have more than one answer.

1. 27	4. 62	7. 1,255	10. 2,170
2. 45	5. 186	8. 64	11. 4,008
3. 50	6. 486	9. 820	12. 63

1.9 Dividing Whole Numbers

Example 12: Divide: $4,993 \div 24$

Step 1: Rewrite the problem using the symbol $\overline{)\quad}$.

Step 2: Divide 24 into 49. Multiply 2×24, and subtract.

$$\begin{array}{r} 2 \\ 24\overline{)4993} \\ -48 \\ \hline 19 \end{array}$$

Step 3: You will notice you cannot divide 24 into 19. You must put a 0 in the answer and then bring down the 3.

$$\begin{array}{r} 20 \\ 24\overline{)4993} \\ -48 \\ \hline 193 \end{array}$$

Step 4: Divide 24 into 193. Multiply 8×24, and subtract.

$$\begin{array}{r} 208 \text{ r } 1 \\ 24\overline{)4993} \\ -48 \\ \hline 193 \\ -192 \\ \hline 1 \end{array}$$

The answer is 208 with a remainder of 1.

Divide.

1. $6,274 \div 13$
2. $2,384 \div 43$
3. $12,747 \div 24$
4. $5,417 \div 19$
5. $8,042 \div 27$
6. $9,548 \div 63$
7. $6,254 \div 41$
8. $4,362 \div 35$
9. $4,345 \div 25$
10. $15,467 \div 43$
11. $7,412 \div 54$
12. $9,379 \div 83$
13. $9,547 \div 31$
14. $7,436 \div 61$
15. $5,464 \div 38$
16. $23,567 \div 11$

1.10 Multiplying and Dividing by Multiples of Ten

Multiples of ten are 10, 100, 1,000, 10,000, and so on. Multiplying by a multiple of ten is simple.

Example 13: Multiply 479×100.

Step 1: Count the number of 0's in the multiple of 10. There are 2 0's in 100.

Step 2: Add two zeros to the other factor in the problem and write the answer.
$479 \times 100 = 47,900$

Find the product.

1. 27×100
2. $356 \times 1,000$
3. $471 \times 100,000$
4. $3,714 \times 100$
5. $2,642 \times 1,000$
6. $1,261 \times 100,000$
7. 39×100
8. $42 \times 1,000$
9. $417 \times 100,000$

Dividing by a multiple of 10 is easy.

Example 14: Divide $2,700 \div 100$

Step 1: Count the number of 0's in the multiple of 10. There are 2 0's in 100.

Step 2: Take two zeros away from the other factor in the problem and write the answer.
$2,700 \div 100 = 27$

Find the quotient.

10. $80,000 \div 10$
11. $100,000 \div 100$
12. $200,000 \div 10$
13. $40,000 \div 1,000$
14. $540,000,000 \div 1,000,000$
15. $27,000 \div 1,000$
16. $35,000 \div 10$
17. $300,000 \div 10,000$
18. $500,000 \div 100$
19. $33,000 \div 1,000$
20. $330,000 \div 10,000$
21. $360,000 \div 10$
22. $400,000,000 \div 10,000,000$
23. $150,000,000 \div 100,000$

1.11 Factors

Factor: A number that is multiplied to get a product.

Example 15: 2 is a factor of 4 because $2 \times 2 = 4$

(the two 2s are labelled "factors"; the 4 is labelled "product")

Example 16: List all factors for the number 16.

Step 1: How many ways can we multiply 2 whole numbers to gain the product 16? Review the chart below.

Factor		Factor		Product
1	×	16	=	16
2	×	8	=	16
4	×	4	=	16
8	×	2	=	16

Step 2: Without duplicating the numbers listed in the chart above, we have 1, 2, 4, 8, 16. These numbers are factors of 16.
The factors of 16 are 1, 2, 4, 8, and 16.

Example 17: List all factors for the number 24.

Factor		Factor		Product
1	×	24	=	24
2	×	12	=	24
3	×	8	=	24
4	×	6	=	24
6	×	4	=	24
8	×	3	=	24
12	×	2	=	24
24	×	1	=	24

Answer: 1, 2, 3, 4, 6, 8, 12, and 24

List all the factors for the numbers below.

1. 6
2. 15
3. 11
4. 27
5. 45
6. 99
7. 100
8. 10
9. 13
10. 14
11. 4
12. 32

1.12 Greatest Common Factor

Example 18: Find the greatest common factor (GCF) of 16 and 24.

To find the **greatest common factor (GCF)** of two numbers, first list the factors of each number.

The factors of 16 are: 1, 2, 4, 8, and 16
The factors of 24 are: 1, 2, 3, 4, 6, 8, 12, and 24
What is the **largest** number they both have in common? **8**
8 is the **greatest** (largest number) **common factor**.

Find all the factors and the greatest common factor (GCF) of each pair of numbers below.

	Pairs	Factors	GCF		Pairs	Factors	GCF
1.	10			9.	6		
	15				42		
2.	12			10.	14		
	16				63		
3.	18			11.	9		
	36				51		
4.	27			12.	18		
	45				45		
5.	32			13.	12		
	40				20		
6.	16			14.	16		
	48				40		
7.	14			15.	10		
	42				45		
8.	4			16.	18		
	26				30		

1.13 Prime and Composite Numbers

A **prime** number is a number greater than 1 that can only be divided by itself and 1 without a remainder.

For example, 17 is a **prime** number.

17 can only be divided by 1 and 17.

1 and 17 are the only **factors** of 17.

A **composite** number is a number greater than 1 that can be divided by 1, itself, and at least one other number.

16 is a **composite** number.

16 can be divided by 1, 2, 4, 8, and 16.

1, 2, 4, 8, and 16 are the **factors** of 16.

List the factors of each number below. Then write P if it is a prime number or C if it is a composite number.

	Number	Factors	Prime or Composite
1.	24	1, 2, 3, 4, 6, 8, 12, 24	C
2.	33		
3.	25		
4.	42		
5.	14		
6.	35		
7.	47		
8.	49		
9.	18		
10.	56		
11.	71		
12.	52		
13.	19		
14.	31		
15.	26		
16.	16		
17.	81		
18.	44		
19.	90		
20.	45		
21.	13		
22.	9		
23.	12		
24.	27		

1.14 Prime Factorization

Prime factorization is the process of factoring a number into prime numbers. A prime number, also called a prime, is a number that can only be divided by itself and 1. There are two main ways of finding the primes of a number: dividing and splitting.

Example 19: Find the primes of 66 by division.

Step 1: To find the primes by division, you must only divide 66 by prime numbers until you can only divide by one.

$66 \div \mathbf{2} = 33$, $33 \div \mathbf{3} = 11$, $11 \div \mathbf{11} = 1$, $11 \div 1 = 11$ (1 is not prime)

Step 2: All the prime numbers used as divisors make up the prime factorization of 6.
$66 = 2 \times 3 \times 11$

Check: To check, multiply the prime numbers together, and you should get the original value.

Example 20: Find the primes of 66 and 120 using the splitting method.

Step 1: In this method, the number must be split by any two factors until all of the factors are prime.

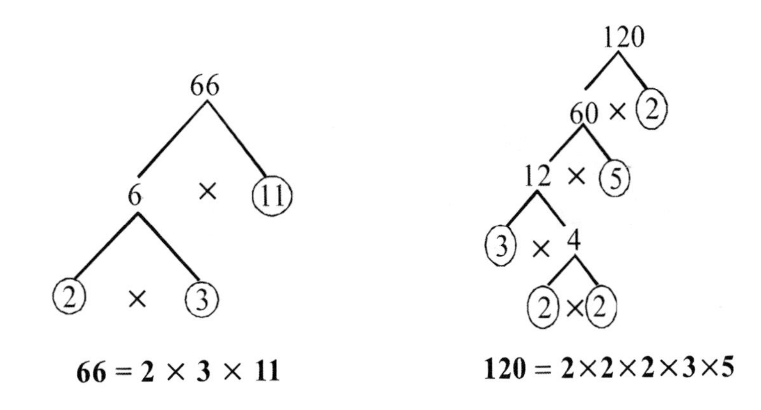

Step 2: All the prime numbers used in splitting 66 make up the prime factorization of 66. All the prime numbers used in splitting 120 make up the prime factorization of 120.

Hint: The factors found during prime factorization should always be prime, should always multiply together to get the correct answer, and should always be listed from least to greatest.

Find the prime factorization of each number using the division method. The first one has been done for you.

1. $10 \div \mathbf{2} = 5 \div \mathbf{5} = 1$
   $10 = 2 \times 5$
2. 14
3. 55
4. 110
5. 126
6. 142
7. 8
8. 21
9. 32
10. 36
11. 51
12. 84
13. 125
14. 48
15. 77
16. 65
17. 200
18. 413

Find the prime factorization of each number using the splitting method. The first one has been done for you.

19.

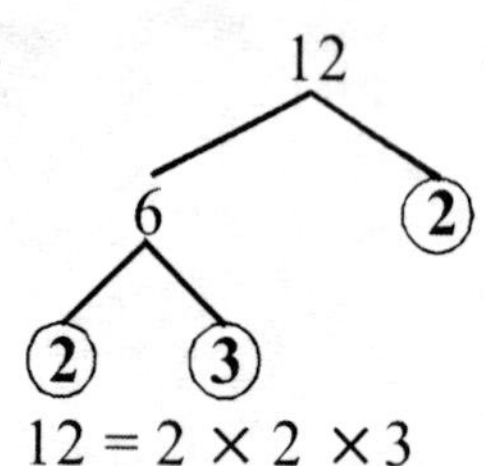

$12 = 2 \times 2 \times 3$

20. 24
21. 45
22. 120
23. 52
24. 91
25. 18
26. 67
27. 20
28. 15
29. 35
30. 122

1.15 Multiples

To find the **multiple** of any whole number, multiply that whole number by 1, then 2, 3, 4, 5, etc.

For example, to find multiples for the number 4, multiply the number 4 by 1, 2, 3, 4, 5, etc. The products 4, 8, 12, 16, and 20 are all multiples of 4.

Example 21: Find multiples of 2.

Step 1: Multiply 2 by 1, 2, 3, 4, 5, etc.

$2 \times 1 = 2$

$2 \times 2 = 4$

$2 \times 3 = 6$

$2 \times 4 = 8$

$2 \times 5 = 10$

Step 2: List the products from step 1: 2, 4, 6, 8, 10

The first five multiples of 2 are 2, 4, 6, 8, and 10.

Example 22: Find multiples of 3.

Step 1: Multiply 3 by 1, 2, 3, 4, 5, etc.

$3 \times 1 = 3$

$3 \times 2 = 6$

$3 \times 3 = 9$

$3 \times 4 = 12$

$3 \times 5 = 15$

Step 2: List the products from step 1: 3, 6, 9, 12, 15

The first five multiples of 3 are 3, 6, 9, 12, and 15.

The chart below demonstrates 5 multiples for the numbers 2, 3, 4, 5.

$\times$	**1**	**2**	**3**	**4**	**5**	**Multiples**
2	2	4	6	8	10	2,4,6,8,10
3	3	6	9	12	15	3,6,9,12,15
4	4	8	12	16	20	4,8,12,16,20
5	5	10	15	20	25	5,10,15,20,25

Example 23: Is 30 a multiple of 6?

Step 1: Multiply 3 by 1, 2, 3, 4, 5, etc. until we get to 30.

$6 \times 1 = 6$

$6 \times 2 = 12$

$6 \times 3 = 18$

$6 \times 4 = 24$

$6 \times 5 = 30$ STOP

Step 2: Looking at the multiples in step 1, we see that $6 \times 5 = 30$, so 30 is a multiple of 6.

Which of the numbers below are multiples of 7?

1. 16
2. 21
3. 35
4. 27
5. 49
6. 11
7. 14
8. 22

9. Is 12 a multiple of 6?

1.16 Least Common Multiple

To find the **least common multiple (LCM)**, of two numbers, first list the multiples of each number. The multiples of a number are 1 times the number, 2 times the number, 3 times the number, and so on.

The multiples of 6 are: 6, 12, 18, 24, 30...

The multiples of 10 are: 10, 20, 30, 40, 50...

What is the smallest multiple they both have in common? 30

30 is the **least** (smallest number) **common multiple** of 6 and 10.

Find the least common multiple (LCM) of each pair of numbers below.

	Pairs	Multiples	LCM		Pairs	Multiples	LCM
1.	6	6, 12, 18, 24, 30	30	10.	6		
	15	15, 30			7		
2.	12			11.	4		
	16				18		
3.	18			12.	7		
	36				5		
4.	7			13.	30		
	3				45		
5.	12			14.	3		
	8				8		
6.	6			15.	12		
	8				9		
7.	4			16.	5		
	14				45		
8.	9			17.	3		
	6				5		
9.	2			18.	4		
	15				22		

1.17 Whole Number Word Problems

1. If Jacob averages 15 points per basketball game, how many points will he score in a season with 12 games?

2. A cashier can ring up 12 items per minute. How long will it take the cashier to ring up a customer with 72 items?

3. Mrs. Randolph has 26 students in 1st period, 32 students in 2nd period, 27 students in 3rd period, and 30 students in 4th period. What is the total number of students Mrs. Randolph teaches?

4. When Gerald started on his trip, his odometer read 109,875. At the end of his trip it read 110, 480. How many miles did he travel?

5. The Beta Club is raising money by selling boxes of candy. It sold 152 boxes on Monday, 236 boxes on Tuesday, 107 boxes on Wednesday, and 93 boxes on Thursday. How many total boxes did the Beta Club sell?

6. Jonah won 1,056 tickets in the arcade. He purchased a pair of binoculars for 964 tickets. How many tickets does he have left?

7. A school cafeteria has 52 tables. If each table seats 14 people, how many people can be seated in the cafeteria?

8. Leadville, Colorado is 14,286 feet above sea level. Denver, Colorado is 5,280 feet above sea level. What is the difference in elevation between these two cities?

9. The local bakery made 288 doughnuts on Friday morning. How many dozen doughnuts did they make?

10. Mattie ate 14 chocolate-covered raisins. Her big brother ate 5 times as many. How many chocolate-covered raisins did her brother eat?

11. Concession stand sales for a football game totaled $1,563. The actual cost for the food and beverages was $395. How much profit did the concession stand make?

12. An orange grove worker can harvest 480 oranges per hour by hand. How many oranges can the worker harvest in an 8-hour day?

1.18 Determining the Operation

In order to solve a problem, determining which operation to use is essential. When reading a problem, certain key words may help you to determine which operation to use.

Which Operation Should I Use?

Key Words/Phrases	**Operation**
In all, In total, Sum of	Addition
Are left, Remaining	Subtraction
Equal installments, Times	Multiplication
In each set	Division

Example 24: Jackie wants to purchase three new video games from her friend, Pam. The cost of each video game is \$39.50. She plans to pay Pam in 6 monthly installments. What will be her monthly installment?

Answer: Compute the total due to Pam: \$39.50 per game times 3 games = \$118.50
Calculate her installment: \$118.50 $\div$ 6 months = \$19.75
Jackie's monthly installment will be \$19.75.

Example 25: Tyler earns \$25.00 per lawn mowed. If he mows 13 lawns per month, how much will Tyler earn?

Answer: \$25.00 per lawn $\times$ 13 lawns = \$325.00
Tyler will earn \$325.00 per month.

Solve the problems.

1. Kayla went to the movies and purchased candy, popcorn, and soda for 2 of her friends. If the total was \$12.50, how much money did she spend on each of her friends? Bonus: If she paid with \$20.00, how much change should she receive?

2. James went to the bookstore and purchased 6 science fiction books and 4 mystery books. If the science-fiction books cost \$4.95 each and the mystery books cost \$4.25 each, how much did he spend in total?

3. Azrina baked 4 cakes for a school bake sale. If each cake had 12 slices that sold for \$0.25 each, how much money did she earn?

4. Keith bought 4 pizza pies for his soccer team. The cost was \$48.60. The 12 teammates plan to reimburse him for the total charge. If they divide the cost evenly, how much will each teammate contribute?

Chapter 1 Review

Write the value of the number 4.

1. 14
2. 4,689,108
3. 12,047,622

Use the number below to answer the following questions.

425,006

Which number is in the

4. Tens place?
5. Ones place?
6. Hundred-thousands place?

Perform the operation.

7. Add: $18 + 694 + 123 + 75$

8. Subtract: $943 - 768$

9. $\begin{array}{r} 4095 \\ +\ 1827 \\ \hline \end{array}$

10. $\begin{array}{r} 36 \\ \times\ 48 \\ \hline \end{array}$

11. Multiply: 452×23

12. Divide: $786 \div 6$

13. $\begin{array}{r} 2006 \\ -\ 1995 \\ \hline \end{array}$

14. $3\overline{)5721}$

Find the greatest common factor for the following pairs of numbers.

15. 9 and 15

16. 12 and 16

17. 10 and 25

18. 8 and 24

Label the following numbers as C for composite or P for prime. List all factors.

19. 15 ____ ______________________

20. 16 ____ ______________________

21. 19 ____ ______________________

22. 20 ____ ______________________

23. 21 ____ ______________________

24. 23 ____ ______________________

25. 25 ____ ______________________

Find the least common multiple for the following pairs of numbers.

26. 8 and 12

27. 5 and 9

28. 4 and 10

29. 6 and 8

Solve the word problems.

30. Tisha bought a skirt for $25, a blouse for $16, and 2 pairs of hose for $2 each. Her sales tax was $2. What was her change from $50.00?

31. Elaina was selling doughnuts for a school fund-raiser. She sold 66 doughnuts and had 32 left. How many doughnuts did she have to begin with?

32. The Bing family's odometer read 65,453 before driving to Disney World for vacation. After their vacation, the odometer read 66,245. How many miles did they drive during their vacation?

Chapter 1 Test

1. What is the place value of the underlined digit?

   <u>4</u>,319,612

   A. millions
   B. ten-thousands
   C. hundreds
   D. hundred-thousands

2. Find the difference.

$$\begin{array}{r} 1008 \\ -\ \ 879 \\ \hline \end{array}$$

   A. 239
   B. 129
   C. 229
   D. 139

3. Find the product. $527 \times 41 =$

   A. 20,607
   B. 21,607
   C. 2,635
   D. 21,507

4. What is the quotient? $282 \div 2 =$

   A. 131
   B. 140
   C. 141
   D. 116

5. Which number is a factor of the number 32?

   A. 14
   B. 5
   C. 7
   D. 16

6. What is the prime factorization for the number 18?

   A. $2 \times 3 \times 3$
   B. 9×2
   C. 6×3
   D. $6 \times 3 \times 1$

7. What is the GCF for the numbers 24 and 32?

   A. 4
   B. 8
   C. 2
   D. 16

8. Which number is a multiple of 4?

   A. 13
   B. 18
   C. 8
   D. 21

9. What is the LCM for the numbers 5 and 6?

   A. 30
   B. 25
   C. 60
   D. 80

10. Subtract: $564 - 182 =$

   A. 746
   B. 446
   C. 282
   D. 382

11. Which pair contains an even and an odd number?

A. 205 and 37
B. 444 and 555
C. 360 and 440
D. 465 and 917

12. Pamela's mother purchased five boxes of candy bars. If each box contains 20 bars, how many candy bars does Pamela's mother have?

A. 4
B. 25
C. 125
D. 100

13. Select the operation needed to solve this problem.
Jason has 120 pieces of candy. He wants to put equal amounts of candy in each of his party bags. If he has 10 party bags in all, how many pieces will he put in each bag?

A. Addition
B. Subtraction
C. Multiplication
D. Division

14. Which of the following numbers is prime?

A. 9
B. 11
C. 12
D. 15

15. Which of the following lists all the factors of the number 12?

A. 1, 12
B. 2, 3, 4, 6
C. 5, 11
D. none of the above

16. Kenya is reading a novel. She read 25 pages on Monday, 32 pages on Tuesday, and 15 pages on Wednesday. How many total pages did she read?

A. 57
B. 62
C. 72
D. 47

17. Patty sold 24 out of a total of 100 boxes of cookies. How many does she have left?

A. 76
B. 86
C. 124
D. 240

18. $\begin{array}{r} 406 \\ \times\ 109 \\ \hline \end{array}$

A. 5,014
B. 7,694
C. 44,254
D. 47,344

19. $7\overline{)11473}$

A. 163
B. 639
C. 1,639
D. 1,740

20. Mrs. Campbell's 5th grade class is going on a field trip. There are 29 children in the class. Parents are driving, and there will be 4 students per car. What is the smallest number of cars they will need for the children?

A. 6
B. 7
C. 8
D. 9

21. Jed has 155 head of cattle. Each eats 31 pounds of silage every day. How much silage does Jed feed his cattle every day?

    A. 5 lb
    B. 3,705 lb
    C. 4,705 lb
    D. 4,805 lb

22. Jerry set up 18 rows of chairs and put 9 chairs in each row. How many chairs did he set up?

    A. 27
    B. 81
    C. 162
    D. 189

23. Eric's mom drove 11 miles each way to bring Eric to school in the morning and back home in the afternoon. How many miles did she drive for 10 days of school?

    A. 110 miles
    B. 220 miles
    C. 330 miles
    D. 440 miles

Chapter 2
Decimals

This chapter covers the following Georgia Performance Standards:

M5N	Numbers and Operations	M5N2.a, b M5N3.a, b, c, d
M5P	Process Skills	M5P1.a, b M5P4.a, b, c

2.1 Place Value: Less Than One

Place Value Chart from 1 to 0.001

Ones		*Tenths*	*Hundredths*	*Thousandths*
1.0		$\frac{1}{10}$ *or* 0.1	$\frac{1}{100}$ *or* 0.01	$\frac{1}{1,000}$ *or* 0.001
1	.	9	8	7

Let's review the place value chart from left to right. The ones column was discussed earlier. Now, let's see what happens when we look at the numbers after the decimal point.

From the left, let us begin with the tenths column. As we move from column to column, we divide by 10. Example 0.1 divided by 10 = 0.01

The number above in standard form: 1.987

Expanded Decimal Form: $1 + 0.9 + 0.08 + 0.007$

Example 1: Review the chart on the previous page. Which number is in the hundredths place?

Answer: 8 is in the hundredths place. The value is calculated as follows:

$8 \times 0.01 = 0.08$

Example 2: Which number is in the tenths place?

Answer: 9 is in the tenths place. The value is calculated as follows.

$9 \times 0.1 = 0.9$

What is the place value of the number 2 in the following decimal numbers?

1. 0.028

2. 0.214

3. 0.002

4. 0.126

5. 0.182

6. 0.532

7. 1.912

8. 0.002

9. 1.2

Review the numbers below and write the place value of the underlined digit.

10. $32.89\underline{3}$

11. $4.0\underline{7}6$

12. $63.86\underline{7}$

13. $1.00\underline{4}$

14. $75.\underline{6}18$

15. $39.88\underline{6}$

2.2 Multiplying Decimal by Multiples of Ten

Multiples of Ten are 10, 100, and 1,000

Example 3: $23 \times 10 = ?$

Step 1: Count the number of zeroes found in 10, 100, 1,000, etc. In this example, 10 only has one zero. We are going to move the decimal place over one place to the right, since 10 only has one zero.

23.0

Step 2: The decimal place is after the 3 even though you don't see it. When you move the decimal place over one to the right, there is nothing there, so we add a zero. 23 becomes 230
Answer: 230

Example 4: $4.5 \times 100 = ?$

Step 1: 100 has two zeroes. We are going to move the decimal place over two places to the right.

4.50

Step 2: When you move the decimal place over two places to the right, you move once to the right after the five and then one more time and add one zero. 4.5 becomes 450
Answer: 450

Example 5: $880 \times 1,000 = ?$

Step 1: 1,000 has 3 zeroes. We are going to move the decimal place over three places to the right.

880.000

Step 1: When you move the decimal place over three to the right, you must put three zeros after 880. You get 880,000.
Answer: 880,000

Find the product.

1. 278×100
2. 135×10
3. $1,876 \times 100$
4. 50.7×1000
5. 865×100
6. 444×1000
7. 9.870×10
8. 14.92×100
9. 16×1000

Multiplying by 0.1 and 0.01

Example 6: $34 \times 0.1 = ?$

Step 1: Count the number of places after the decimal point. In this example, there is ONE decimal place.

Step 2: Move the decimal in the other factor to the left, as many places as counted in step 1.
Move the decimal to the left ONE place.
34.

Step 3: Answer: 3.4

Example 7: $86 \times 0.01 = ?$

Step 1: 0.01 has two places after the decimal point, including the number one.

Step 2: Move the decimal in 86, two spaces to the left.
86.

Step 3: Answer: 0.86 (Remember to put a zero in the one's place, left of the decimal.)

Note: When moving the decimal, insert zeroes if you no longer have enough digits.

Example 8: $5 \times 0.01 = ?$

Step 1: 0.01 has 2 places after the decimal point, including the number one.

Step 2: Move decimal 2 spaces to the left. 5 becomes 0.05
05.

Step 3: Answer: 0.05

Find the product.

1. 851×0.1
2. 65×0.01
3. 8×0.01
4. 11×0.1
5. 362×0.01
6. 556×0.1
7. 1789×0.01
8. 457×0.1
9. 3×0.01

2.3 Multiplying Decimals by Whole Numbers

Example 9: 87×2.4

Step 1: If written horizontally, rewrite the problem vertically.

$$\begin{array}{r} 87 \\ \times \quad 2.4 \\ \hline \end{array}$$

Step 2: Multiply as if all numbers are whole. Disregard the decimal point, for now.

$$\begin{array}{r} 87 \\ \times \quad 2.4 \\ \hline 348 \\ + \quad 1740 \\ \hline 2088 \end{array}$$

Step 3: For each factor, count the numbers that appear after the decimal point.

87	0 numbers after the decimal
2.4	+ 1 number after the decimal
	1 number after the decimal in total

Step 4: If there is one number after the decimal point in the problem, there should be one number after the decimal point in the answer.
Product should have one (0 + 1) number after the decimal: 208.8

Answer: 208.8

Multiply.

1. 6.7×8
2. 18.9×7
3. 21.1×4
4. $1,122 \times 1.2$
5. 85.8×9
6. 48×3.41
7. 12×7.9
8. 54×66.51
9. 95×14.2

2.4 Multiplying Decimals by Decimals

Example 10: 56.2×0.17

Step 1: Set up the problem as if you were multiplying whole numbers.

$$\begin{array}{r} 56.2 \\ \times\ 0.17 \\ \hline \end{array}$$

Step 2: Multiply as if you were multiplying whole numbers.

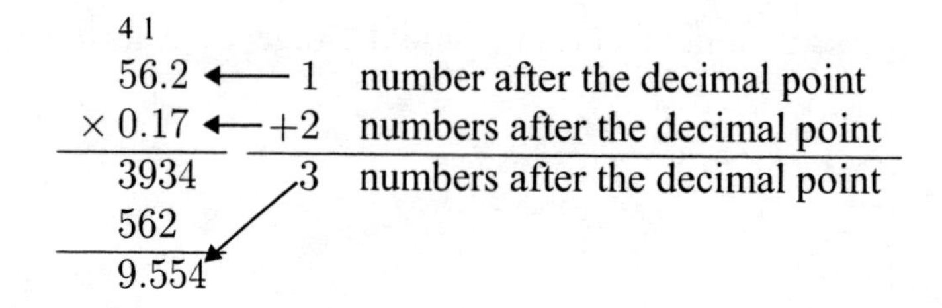

Step 3: Count how many numbers are after the decimal points in the problem. In this problem, three numbers, 2, 1, and 7, come after decimal points, so the answer must also have three numbers after the decimal point.
Answer: 9.554

Multiply.

1. 15.2×3.5
2. 9.54×5.3
3. 5.72×6.3
4. 4.8×3.2
5. 45.8×2.2
6. 4.5×7.1
7. 0.052×0.33
8. 4.12×6.8
9. 23.65×9.2
10. 1.54×0.43
11. 0.47×6.1
12. 1.3×1.57
13. 16.4×0.5
14. 0.87×3.21
15. 5.94×0.65
16. 7.8×0.23

2.5 More Multiplying Decimals by Decimals

Example 11: Find 0.007×0.125

Step 1: Multiply as you would whole numbers.

$$\begin{array}{r} 0.007 \\ \times\ 0.125 \\ \hline 0.000875 \end{array} \begin{array}{l} \leftarrow 3 \text{ numbers after the decimal point} \\ \leftarrow +3 \text{ numbers after the decimal point} \\ \hline \leftarrow 6 \text{ numbers after the decimal point} \end{array}$$

Step 2: Count how many numbers are behind decimal points in the problem. In this case, six numbers come after decimal points in the problem, so there must be six numbers after the decimal point in the answer. In this problem, 0's needed to be written in the answer in front of the 8, so there would be 6 numbers after the decimal point.
Answer: 0.000875

Multiply. Write in zeroes as needed. Round dollar figures to the nearest penny.

1. $0.123 \times .45$
2. 0.004×10.31
3. 1.54×1.1
4. 10.05×0.45
5. 9.45×0.8
6. $\$6.49 \times 0.06$
7. 5.003×0.009
8. $\$9.99 \times 0.06$
9. 6.09×5.3
10. $\$22.00 \times 0.075$
11. 5.914×0.02
12. 4.96×0.23
13. 6.98×0.02
14. 3.12×0.08
15. 7.158×0.09
16. 0.0158×0.32

2.6 Division of Decimals by Whole Numbers

Example 12: $52.26 \div 6$

Step 1: Copy the problem as you would for whole numbers. Copy the decimal point directly above in the place for the answer.

$$6 \overline{)52.26}$$

Step 2: Divide the same way as you would with whole numbers.

$$\begin{array}{r} 8.71 \\ 6 \overline{)52.26} \\ 48 \\ 4\ 2 \\ -\ 4\ 2 \\ \hline 6 \\ -\ 6 \\ \hline 0 \end{array}$$

Divide. Remember to copy the decimal point directly above the place for the answer.

1. $42.75 \div 3$
2. $74.16 \div 6$
3. $81.50 \div 25$
4. $82.46 \div 14$
5. $12.50 \div 2$
6. $224.64 \div 52$
7. $183.04 \div 52$
8. $281.52 \div 23$
9. $72.36 \div 4$
10. $379.5 \div 15$
11. $152.25 \div 21$
12. $40.375 \div 19$
13. $102.5 \div 5$
14. $113.4 \div 9$
15. $585.14 \div 34$
16. $93.6 \div 24$

2.7 Division of Decimals by Decimals

Example 13: $374.5 \div 0.07$

Step 1: Copy the problem as you would for whole numbers.

$0.07 \overline{)374.5}$ (Divisor: 0.07; Dividend: 374.5)

Step 2: You cannot divide by a decimal number. You must move the decimal point in the divisor 2 places to the right to make it a whole number. The decimal point in the dividend must also move to the right the same number of places. Notice that in this example, you must add a 0 to the dividend.

$0.07. \overline{)374.50.}$

Step 3: The problem now becomes $37,450 \div 7$. Copy the decimal point from the dividend straight above in the place for the answer.

$$
\begin{array}{r}
5350. \\
7 \overline{)37450.} \\
-35 \\
\hline
24 \\
-21 \\
\hline
35 \\
-35 \\
\hline
00
\end{array}
$$

Divide. Remember to move the decimal points.

1. $0.676 \div 0.013$
2. $70.32 \div 0.08$
3. $\$54.60 \div 0.84$
4. $\$10.35 \div 0.45$
5. $18.46 \div 1.3$
6. $14.6 \div 0.002$
7. $\$125.25 \div 0.75$
8. $\$33.00 \div 1.65$
9. $154.08 \div 1.8$
10. $0.4374 \div 0.003$
11. $292.9 \div 0.29$
12. $6.375 \div 0.3$
13. $4.8 \div 0.08$
14. $1.2 \div 0.024$
15. $15.725 \div 3.7$
16. $\$167.50 \div 0.25$

2.8 Ordering Decimals

Example 14: Order the following decimals from greatest to least.

0.3, 0.029, 0.208, 0.34

Step 1: Arrange numbers with decimal points directly under each other.

0.3
0.029
0.208
0.34

Step 2: Fill in with zeros so they all have the same number of places after the decimal point. Remember to **read the numbers as if the decimal points were not there.**

0.300
0.029 ← Least
0.208
0.340 ← Greatest

Answer: 0.34, 0.3, 0.208, 0.029

Order each set of decimals below from greatest to least.

1. 0.075, 0.705, 0.7, 0.75
2. 0.5, 0.56, 0.65, 0.06
3. 0.9, 0.09, 0.099, 0.95
4. 0.6, 0.59, 0.06, 0.66
5. 0.3, 0.303, 0.03, 0.33
6. 0.02, 0.25, 0.205, 0.5
7. 0.004, 0.44, 0.045, 0.4
8. 0.59, 0.905, 0.509, 0.099
9. 0.1, 0.01, 0.11, 0.111
10. 0.87, 0.078, 0.78, 0.8
11. 0.41, 0.45, 0.409, 0.49
12. 0.754, 0.7, 0.74, 0.75
13. 0.63, 0.069, 0.07, 0.06
14. 0.23, 0.275, 0.208, 0.027

Order each set of decimals below from least to greatest.

15. 0.055, 0.5, 0.59, 0.05
16. 0.7, 0.732, 0.74, 0.72
17. 0.04, 0.48, 0.048, 0.408
18. 0.9, 0.905, 0.95, 0.09
19. 0.19, 0.09, 0.9, 0.1
20. 0.21, 0.02, 0.021, 0.2

2.9 Decimal Word Problems

1. Micah can have his bike fixed for \$19.99, or he can buy the new part for his bike and replace it himself for \$8.79. How much would he save by fixing his bike himself?

2. Megan buys 5 boxes of cookies for \$3.75 each. What is her change from \$20.00?

3. Will subscribes to a monthly sports magazine. His one-year subscription costs \$29.97. If he pays for the subscription in 3 equal installments, how much is each payment?

4. Pat purchases 2.5 pounds of jelly beans at \$0.98 per pound. What is his change from \$10.00?

5. The White family took \$650 cash with them on vacation. At the end of their vacation, they had \$4.67 left. How much cash did they spend on vacation?

6. Acer Middle School spends \$1, 443.20 on 55 math books. How much does each book cost?

7. The Junior Beta Club needs to raise \$1, 513.75 to go to a national convention. The parents donated \$850.00. If they decide to sell candy bars at \$1.25 each to earn the rest of the money, how many must they sell to meet their goal?

8. Fleta owns a candy store. On Monday, she sold 6.5 pounds of chocolate, 8.34 pounds of jelly beans, 4.9 pounds of sour snaps, and 5.64 pounds of yogurt-covered raisins. How many pounds of candy did she sell in total?

9. Randal purchased a rare coin collection for \$1, 803.95. He sold it at auction for \$2, 700. How much money did he make on the coins?

10. A leather jacket that normally sells for \$259.99 is on sale now for \$197.88. How much can you save if you buy it now?

Chapter 2 Review

Review the numbers below and write the place value of the underlined digit.

1. $57.18\underline{4}$
2. $69.4\underline{3}1$
3. $1.\underline{7}$

Find the product.

4. 450.6×10
5. 19.585×100
6. $16,554 \times 100$
7. 524×0.1
8. 856×0.01
9. 23×0.01

Multiply.

10. 108×18.4
11. 88×14.2
12. 65×9.5
13. 4.58×0.025
14. 0.879×1.7
15. 30.7×0.0041

Divide.

16. $17.28 \div 0.054$
17. $174.66 \div 1.23$
18. $2.115 \div 9$

Solve the word problems below.

19. Gene works for his father sanding wooden rocking chairs. He earns \$6.35 per chair. How many chairs does he need to sand in order to buy a portable radio/CD player for \$146.05?

20. Margo's Mint Shop has a machine that produces 4.35 pounds of mints per hour. How many pounds of mints are produced in each 8-hour shift?

21. Carter's Junior High track team runs the first leg of a 400-meter relay race in 10.23 seconds, the second leg in 11.4 seconds, the third leg in 10.77 seconds, and the last leg in 9.9 seconds. How long does it take for them to complete the race?

Chapter 2 Test

1. What is the place value of the number 4 in the number 9.2<u>4</u>6?

   A. Four hundred
   B. Four hundredths
   C. Four thousand
   D. Four ones

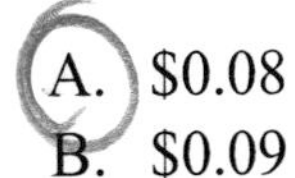

2. What is the place value of the number 6 in the number 5.14<u>6</u>?

   A. six thousandths
   B. sixty thousandths
   C. six-hundred thousandths
   D. six-hundred thousand

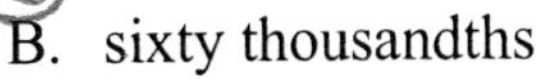

3. Multiply: $487 \times 2.2 =$

   A. 107.41
   B. 107.14
   C. 1,071.4
   D. 1,071.14

4. Multiply: $6.27 \times 0.43 =$

   A. 2.5961
   B. 2.6971
   C. 2.9661
   D. 2.6961

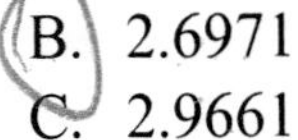

5. Hanna bought 3 pairs of socks priced at 3 for \$5.00 and shoes for \$45.95. She paid \$2.55 sales tax. How much change did she receive from \$100.00?

   A. \$36.50
   B. \$46.50
   C. \$51.50
   D. \$53.50

6. If 15 pencils cost \$1.20, what is the cost of one pencil?

   A. \$0.08
   B. \$0.09
   C. \$0.15
   D. \$0.18

7. Divide: $124 \div 6.2 =$

   A. 20.12
   B. 20
   C. 20.22
   D. 20.02

8. Divide: $16.8 \div 2.1 =$

   A. 8.8
   B. 8.0
   C. 8.4
   D. 8.2

9. Christian earned \$177.00 washing cars this weekend. If he charged \$14.75 per car, how many cars did he wash?

   A. 12
   B. 13
   C. 15
   D. 20

10. John earns \$5.25/hour working as a lifeguard. If he works 25 hours per week, how much will he earn each week?

   A. \$125.25
   B. \$131.75
   C. \$125.50
   D. \$131.25

11. Richard buys a camcorder for $229.95, which is now on sale for $207.99. How much could he have saved if he waited to buy the camcorder on sale?

A. $21.04
B. $21.96
C. $22.06
D. $22.96

12. Tamara bought 2 movie tickets for $6.50 each, 2 colas for $1.25 each, and two bags of popcorn for $1.50 each. How much did Tamara spend in all?

A. $9.25
B. $15.50
C. $18.50
D. $18.75

13. Rosa worked 28 hours this week and was paid $5.60 per hour. What were her total earnings for the week?

A. $15.68
B. $22.40
C. $33.60
D. $156.80

14. Carl needs $231.75 to buy a new bike. If he earns $5.15 per hour net at a part-time job, how many hours would he have to work to earn enough to buy the bike?

A. 43 hours
B. 44.25 hours
C. 45 hours
D. 46.2 hours

15. Seth bought a CD for $13.95 and 2 tapes for $7.99 each. He paid $1.51 sales tax. What was his change from a $50.00 bill?

A. $18.56
B. $26.55
C. $18.49
D. $20.07

16. Patrick bought 2 pens for $0.49 each, one notebook for $1.53, and a protractor for $1.09. He paid $0.16 sales tax. What was his change from $10.00?

A. $6.89
B. $6.73
C. $6.24
D. $6.38

17. Selena bought a wallet for $15.70 and 4 bottles of nail polish for $2.45 each. She paid $1.12 sales tax. What was her change from $40.00?

A. $20.73
B. $14.50
C. $22.57
D. $13.38

18. Kent bought 4 CDs for $13.95 each. What was his change from $60.00?

A. $4.20
B. $46.05
C. $4.00
D. $56.51

19. If you travel 35 miles per hour for 4.5 hours, how far will you travel?

A. 157.5 miles
B. 202.5 miles
C. 140.5 miles
D. 189.0 miles

20. Grandma made a candy cane 39.75 inches long. If she divides it equally among her 3 grandchildren, how long will each piece be?

A. 11.6 inches
B. 10.25 inches
C. 13.25 inches
D. 12.2 inches

Chapter 3
Fractions

This chapter covers the following Georgia Performance Standards:

M5N	Numbers and Operations	M5N4.a, b, c, d, e M5N4.f, g, h, i
M5P	Process Skills	M5P1.a, b M5P4.a, b, c

3.1 Simplifying Fractions

Example 1: Reduce $\frac{4}{8}$ to lowest terms.

Step 1: First you need to find the greatest common factor of 4 and 8. Think: What is the largest number that can be divided into 4 and 8 without a remainder?

These must be the same number. $?\overline{)4}$ $?\overline{)8}$ 4 and 8 can both be divided by 4.

Step 2: Divide the top and bottom of the fraction by the same number.

$\frac{4 \div 4}{8 \div 4} = \frac{1}{2}$ Therefore, $\frac{4}{8} = \frac{1}{2}$.

Simplify the following fractions.

1. $\frac{2}{8}$ 6. $\frac{27}{54}$ 11. $\frac{30}{45}$ 16. $\frac{12}{36}$ 21. $1\frac{2}{12} = 1\frac{1}{6}$ 26. $9\frac{6}{27}$

2. $\frac{12}{15}$ 7. $\frac{14}{22}$ 12. $\frac{16}{64}$ 17. $\frac{13}{39}$ 22. $3\frac{5}{15}$ 27. $15\frac{4}{18}$

3. $\frac{9}{27}$ 8. $\frac{9}{21}$ 13. $\frac{10}{25}$ 18. $\frac{28}{49}$ 23. $4\frac{9}{15}$ 28. $8\frac{8}{28}$

4. $\frac{12}{42}$ 9. $\frac{4}{14}$ 14. $\frac{3}{12}$ 19. $\frac{8}{18}$ 24. $7\frac{24}{48}$ 29. $5\frac{14}{42}$

5. $\frac{3}{21}$ 10. $\frac{6}{26}$ 15. $\frac{15}{30}$ 20. $\frac{14}{21}$ 25. $8\frac{3}{18}$ 30. $10\frac{18}{36}$

3.2 Simplifying Improper Fractions

Example 2: Simplify $\frac{21}{4}$.

$\frac{21}{4} = 21 \div 4 = 5$ remainder 1

The quotient, 5, becomes the whole number portion of the mixed number.

$\frac{21}{4} = 5\frac{1}{4}$ ↰ The remainder, 1, becomes the top number of the fraction.

The bottom number of the fraction always remains the same.

Example 3: Simplify $\frac{11}{6}$.

Step 1: $\frac{11}{6}$ is the same as $11 \div 6$. $11 \div 6 = 1$ with a remainder of 5.

Step 2: Rewrite as a whole number with a fraction. $1\frac{5}{6}$

Simplify the following improper fractions.

1. $\frac{13}{5} =$ ______
2. $\frac{11}{3} =$ ______
3. $\frac{24}{6} =$ ______
4. $\frac{7}{6} =$ ______
5. $\frac{19}{6} =$ ______
6. $\frac{16}{7} =$ ______
7. $\frac{13}{8} =$ ______
8. $\frac{9}{5} =$ ______
9. $\frac{22}{3} =$ ______
10. $\frac{13}{4} =$ ______
11. $\frac{15}{2} =$ ______
12. $\frac{22}{9} =$ ______
13. $\frac{17}{9} =$ ______
14. $\frac{27}{8} =$ ______
15. $\frac{32}{7} =$ ______
16. $\frac{3}{2} =$ ______
17. $\frac{7}{4} =$ ______
18. $\frac{21}{10} =$ ______
19. $6\frac{12}{8} =$ ______
20. $11\frac{7}{5} =$ ______
21. $9\frac{10}{5} =$ ______
22. $2\frac{3}{2} =$ ______
23. $5\frac{11}{6} =$ ______
24. $8\frac{15}{6} =$ ______

3.3 Finding Numerators

Remember, any fraction that has the same non-zero numerator (top numbers) and denominator (bottom number) equals 1.

Example 4: $\frac{5}{5} = 1$ $\quad \frac{8}{8} = 1$ $\quad \frac{12}{12} = 1$ $\quad \frac{15}{15} = 1$ $\quad \frac{25}{25} = 1$

Any fraction multiplied by 1 in any form remains equal to itself.

Example 5: $\frac{3}{7} \times \frac{4}{4} = \frac{12}{28}$ so $\frac{3}{7} = \frac{12}{28}$

Find the missing numerator (top number) $\frac{5}{8} = \frac{}{24}$

Step 1: Ask yourself, "What was 8 multiplied by to get 24?" 3 is the answer.

Step 2: The only way to keep the fraction equal is to multiply the top and bottom number by the same number. The bottom number was multiplied by 3, so multiply the top number by 3, as shown below.

$\frac{5}{8} \times \frac{3}{3} = \frac{15}{24}$ Note: $\frac{3}{3} = 1$

Find the missing numerators from the following equivalent fractions.

1. $\frac{2}{6} = \frac{6}{18}$
2. $\frac{2}{3} = \frac{}{27}$
3. $\frac{4}{9} = \frac{}{18}$
4. $\frac{7}{15} = \frac{}{45}$
5. $\frac{9}{10} = \frac{}{50}$
6. $\frac{5}{6} = \frac{}{36}$
7. $\frac{1}{4} = \frac{}{36}$
8. $\frac{3}{14} = \frac{}{28}$
9. $\frac{2}{5} = \frac{}{25}$
10. $\frac{4}{11} = \frac{}{33}$
11. $\frac{5}{6} = \frac{}{18}$
12. $\frac{6}{11} = \frac{}{22}$
13. $\frac{8}{15} = \frac{}{45}$
14. $\frac{1}{9} = \frac{}{18}$
15. $\frac{7}{8} = \frac{35}{40}$
16. $\frac{1}{12} = \frac{}{48}$
17. $\frac{3}{8} = \frac{}{24}$
18. $\frac{3}{4} = \frac{}{16}$
19. $\frac{2}{7} = \frac{}{49}$
20. $\frac{11}{12} = \frac{22}{24}$
21. $1\frac{2}{5} = 1\frac{}{45}$
22. $2\frac{4}{5} = 2\frac{}{15}$
23. $8\frac{1}{9} = 8\frac{}{27}$
24. $3\frac{3}{8} = 3\frac{}{56}$
25. $11\frac{2}{13} = 11\frac{}{26}$
26. $7\frac{1}{7} = 7\frac{}{35}$
27. $2\frac{4}{5} = 2\frac{}{10}$
28. $1\frac{3}{10} = 1\frac{}{40}$
29. $10\frac{7}{8} = 10\frac{}{48}$
30. $5\frac{6}{7} = 5\frac{}{14}$

3.4 Adding Fractions

Example 6: Add $3\frac{1}{2} + 2\frac{2}{3}$

Step 1: Fractions need the same denominator before they can be added. Rewrite the problem vertically, and find a common denominator. Think: What is the smallest number I can divide 2 and 3 into without a remainder? 6, of course.

$$\begin{array}{rcl} 3\frac{1}{2} & = & \frac{}{6} \\ +2\frac{2}{3} & = & \frac{}{6} \\ \hline \end{array}$$

Step 2: Find the numerators of the fractions as you did on the page before this one.

Step 3: Add whole numbers and fractions, and simplify.

$$\begin{array}{rclcl} 3\frac{1}{2} & = & 3\frac{3}{6} & & \\ +2\frac{2}{3} & = & 2\frac{4}{6} & & \\ \hline & = & 5\frac{7}{6} & = & 6\frac{1}{6} \end{array}$$

Add and simplify the answers.

1. $3\frac{5}{9} + 5\frac{2}{3}$
3. $3\frac{3}{4} + 2\frac{3}{5}$
5. $6\frac{5}{6} + 4\frac{1}{3}$
7. $\frac{1}{3} + 7\frac{3}{4}$
9. $4\frac{7}{10} + 8\frac{2}{3}$
11. $3\frac{3}{11} + 2\frac{3}{4}$
2. $1\frac{1}{4} + 4\frac{2}{5}$
4. $2\frac{1}{4} + 1\frac{7}{8}$
6. $9\frac{1}{5} + 5\frac{5}{6}$
8. $9\frac{4}{9} + 3\frac{2}{3}$
10. $5\frac{2}{7} + \frac{1}{2}$
12. $\frac{3}{5} + \frac{4}{9}$

3.5 Subtracting Fractions

Example 7: Subtract $\frac{7}{8} - \frac{1}{5}$

Step 1: Rewrite vertically.

$$\begin{array}{r} \frac{7}{8} \\ -\frac{1}{5} \\ \hline \end{array}$$

Step 2: Find the common denominator.

$$\begin{array}{rcl} \frac{7}{8} & = & \frac{\ }{40} \\ -\frac{1}{5} & = & \frac{\ }{40} \\ \hline \end{array}$$

Step 3: Find the numerators.

$$\begin{array}{rcl} \frac{7}{8} & = & \frac{35}{40} \\ -\frac{1}{5} & = & \frac{8}{40} \\ \hline \end{array}$$

Step 4: Subtract.

$$\begin{array}{rcl} \frac{7}{8} & = & \frac{35}{40} \\ -\frac{1}{5} & = & \frac{8}{40} \\ \hline & & \frac{27}{40} \end{array}$$

Subtract.

1. $\frac{8}{9} - \frac{3}{4}$
2. $\frac{9}{10} - \frac{2}{3}$
3. $\frac{7}{8} - \frac{4}{5}$
4. $\frac{3}{4} - \frac{2}{5}$
5. $\frac{9}{10} - \frac{1}{4}$
6. $\frac{5}{6} - \frac{1}{4}$
7. $\frac{4}{5} - \frac{2}{3}$
8. $\frac{5}{9} - \frac{1}{3}$

3.6 Subtracting Mixed Numbers from Whole Numbers

Example 8: Subtract $15 - 3\frac{3}{4}$

Step 1: Rewrite the problem vertically.

$$\begin{array}{r} 15 \\ -\quad 3\frac{3}{4} \\ \hline \end{array}$$

Step 2: You cannot subtract three-fourths from nothing. You must borrow 1 from 15. You will need to put the 1 in the fraction form. If you use $\frac{4}{4}$ $\left(\frac{4}{4} = 1\right)$, you will be ready to subtract.

$$\begin{array}{r} \overset{4}{1\not{5}}\frac{4}{4} \\ -\quad 3\frac{3}{4} \\ \hline 11\frac{1}{4} \end{array}$$

Subtract.

1. $12 - 3\frac{2}{9}$
2. $3 - 1\frac{4}{7}$
3. $24 - 11\frac{4}{5}$
4. $2 - 1\frac{2}{5}$
5. $4 - 1\frac{5}{8}$
6. $11 - 9\frac{7}{8}$
7. $14 - 9\frac{7}{12}$
8. $8 - 3\frac{1}{3}$
9. $5 - 3\frac{1}{2}$
10. $17 - 13\frac{1}{5}$
11. $3 - 1\frac{5}{11}$
12. $13 - 8\frac{9}{10}$
13. $15 - 6\frac{3}{4}$
14. $6 - 4\frac{8}{9}$
15. $20 - 12\frac{6}{7}$
16. $21 - 1\frac{3}{20}$
17. $9 - 5\frac{2}{3}$
18. $8 - 7\frac{3}{5}$
19. $5 - 4\frac{5}{8}$
20. $14 - 9\frac{1}{7}$
21. $12 - 4\frac{1}{6}$
22. $2 - 1\frac{2}{3}$
23. $42 - 30\frac{2}{9}$
24. $7 - 5\frac{9}{13}$
25. $19 - 13\frac{3}{8}$
26. $14 - 10\frac{5}{9}$
27. $16 - 8\frac{1}{4}$
28. $15 - 3\frac{5}{7}$

3.7 Subtracting Mixed Numbers with Borrowing

Example 9: Subtract $7\frac{1}{4} - 5\frac{5}{6}$

Step 1: Rewrite the problem and find a common denominator.

$$\begin{array}{rcr} 7\frac{1}{4} \; \frac{\times 3}{\times 3} & \rightarrow & 7\frac{3}{12} \\ -5\frac{5}{6} \; \frac{\times 2}{\times 2} & \rightarrow & -5\frac{10}{12} \\ \hline \end{array}$$

Step 2: You cannot subtract 10 from 3. You must borrow 1 from the 7. The 1 will be in the fraction form $\frac{12}{12}$ which you must add to the $\frac{3}{12}$ you already have, making $\frac{15}{12}$. Subtract whole numbers and simplify.

$$\begin{array}{r} \overset{6}{\cancel{7}}\frac{\overset{15}{\cancel{3}}}{12} \\ -5\frac{10}{12} \\ \hline 1\frac{5}{12} \end{array}$$

Subtract and simplify.

1. $4\frac{1}{3} - 1\frac{5}{9}$
2. $3\frac{4}{9} - 2\frac{5}{6}$
3. $8\frac{4}{7} - 5\frac{1}{3}$
4. $5\frac{2}{5} - 3\frac{1}{2}$
5. $8\frac{2}{5} - 5\frac{3}{10}$
6. $9\frac{2}{5} - 4\frac{3}{4}$
7. $9\frac{3}{4} - 2\frac{1}{3}$
8. $5\frac{1}{7} - \frac{2}{3}$
9. $6\frac{1}{5} - 3\frac{3}{8}$
10. $6\frac{5}{6} - 3\frac{4}{5}$
11. $2\frac{2}{9} - 1\frac{3}{4}$
12. $4\frac{7}{10} - 3\frac{1}{3}$
13. $7\frac{3}{5} - 4\frac{5}{6}$
14. $9\frac{3}{8} - 5\frac{1}{2}$
15. $8\frac{1}{9} - 5\frac{1}{3}$
16. $5\frac{1}{6} - 1\frac{2}{3}$
17. $6\frac{5}{6} - 3\frac{1}{3}$
18. $7\frac{2}{3} - 3\frac{5}{6}$
19. $8\frac{4}{7} - 4\frac{3}{4}$
20. $9\frac{3}{4} - 1\frac{1}{5}$

3.8 Multiplying Fractions with Canceling

Example 10: Multiply $\frac{1}{2} \times \frac{3}{4}$

Step 1: Multiply the top numbers $1 \times 3 = 3$. The top of the fraction is 3.

Step 2: Multiply the bottom numbers $2 \times 4 = 8$. The bottom of the fraction is 8.

$$\frac{1}{2} \times \frac{3}{4} = \frac{3}{8}$$

Multiply the fractions.

1. $\frac{4}{7} \times \frac{3}{5}$
2. $\frac{3}{4} \times \frac{1}{5}$
3. $\frac{2}{3} \times \frac{1}{7}$
4. $\frac{2}{3} \times \frac{1}{3}$
5. $\frac{1}{5} \times \frac{4}{9}$
6. $\frac{3}{10} \times \frac{1}{4}$
7. $\frac{1}{2} \times \frac{1}{3}$
8. $\frac{3}{4} \times \frac{1}{3}$
9. $\frac{1}{9} \times \frac{1}{4}$

Example 11: Multiply $\frac{2}{3} \times \frac{3}{4}$

Step 1: In this problem, the 3's are both divisible by 3, so they cancel. $3 \div 3 = 1$

$$\frac{2}{\cancel{3}_1} \times \frac{\cancel{3}^1}{4} = \frac{2}{1} \times \frac{1}{4}$$

Step 2: The 2 and the 4 are both divisible by 2, so they cancel. $2 \div 2 = 1$ and $4 \div 2 = 2$

$$\frac{\cancel{2}^1}{1} \times \frac{1}{\cancel{4}_2} = \frac{1}{1} \times \frac{1}{2}$$

Step 3: Multiply. $\frac{1}{1} \times \frac{1}{2} = \frac{1}{2}$

Cancel where possible in the following problems, then multiply.

10. $\frac{2}{3} \times \frac{3}{8}$
11. $\frac{3}{4} \times \frac{4}{9}$
12. $\frac{3}{8} \times \frac{2}{3}$
13. $\frac{5}{6} \times \frac{3}{10}$
14. $\frac{2}{7} \times \frac{1}{4}$
15. $\frac{5}{9} \times \frac{1}{5}$
16. $\frac{8}{9} \times \frac{3}{4}$
17. $\frac{9}{10} \times \frac{5}{6}$
18. $\frac{3}{7} \times \frac{7}{9}$
19. $\frac{6}{7} \times \frac{1}{12}$
20. $\frac{2}{5} \times \frac{5}{6}$
21. $\frac{4}{9} \times \frac{3}{8}$

3.9 Changing Mixed Numbers to Improper Fractions

Example 12: Change $4\frac{3}{5}$ to an improper fraction.

Step 1: Multiply the whole number (4) by the bottom number of the fraction (5). $4 \times 5 = 20$

Step 2: Add the top number to the product from Step 1: $20 + 3 = 23$

Step 3: Put the answer from step 2 over the bottom number (5).

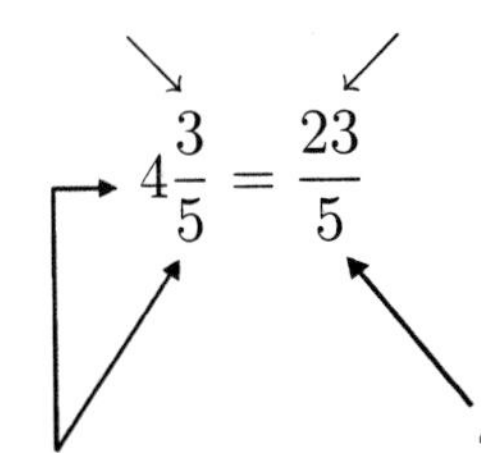

Change the following mixed numbers to improper fractions.

1. $3\frac{1}{2} =$ ______
2. $2\frac{7}{8} =$ ______
3. $9\frac{2}{3} =$ ______
4. $4\frac{3}{5} =$ ______
5. $7\frac{1}{4} =$ ______
6. $8\frac{5}{8} =$ ______
7. $1\frac{2}{7} =$ ______
8. $2\frac{4}{9} =$ ______
9. $6\frac{1}{5} =$ ______
10. $5\frac{2}{7} =$ ______
11. $3\frac{3}{5} =$ ______
12. $9\frac{3}{8} =$ ______
13. $10\frac{4}{5} =$ ______
14. $3\frac{3}{10} =$ ______
15. $4\frac{1}{7} =$ ______
16. $2\frac{5}{6} =$ ______
17. $7\frac{3}{7} =$ ______
18. $6\frac{7}{9} =$ ______
19. $7\frac{2}{5} =$ ______
20. $1\frac{6}{7} =$ ______

Whole numbers become improper fractions when you put them over 1. Change the following whole numbers to improper fractions. The first one is done for you.

21. $4 = \frac{4}{1}$
22. $10 =$ ______
23. $3 =$ ______
24. $2 =$ ______
25. $15 =$ ______
26. $5 =$ ______
27. $6 =$ ______
28. $11 =$ ______
29. $8 =$ ______
30. $16 =$ ______

3.10 Multiplying Mixed Numbers

Example 13: Multiply $4\frac{3}{8} \times \frac{8}{10}$

Step 1: Change the mixed numbers in the problem to improper fractions. $4\frac{3}{8} = \frac{35}{8}$

Step 2: When multiplying fractions, you can cancel and simplify terms that have a common factor. The 8 in the first fraction will cancel with the 8 in the second fraction.

$$\frac{35}{\cancel{8}_1} \times \frac{\cancel{8}^1}{10}$$

The terms 35 and 10 are both divisible by 5, so

35 simplifies to 7, and 10 simplifies to 2.

$$\frac{\cancel{35}^7}{1} \times \frac{1}{\cancel{10}_2} = \frac{7 \times 1}{1 \times 2} = \frac{7}{2}$$

Step 3: Simplify the improper fraction. $\frac{7}{2}$

Step 4: You cannot leave an improper fraction as the answer, so to change $\frac{7}{2}$ back to a mixed number.

$\frac{7}{2} = 3\frac{1}{2}$

Multiply and reduce your answers to lowest terms.

1. $3\frac{1}{5} \times 1\frac{1}{2}$
2. $\frac{3}{8} \times 3\frac{3}{7}$
3. $4\frac{1}{3} \times 2\frac{1}{4}$
4. $4\frac{2}{3} \times 3\frac{3}{4}$
5. $1\frac{1}{2} \times 1\frac{2}{5}$
6. $3\frac{3}{7} \times \frac{5}{6}$
7. $3 \times 6\frac{1}{3}$
8. $1\frac{1}{6} \times 8$
9. $6\frac{2}{5} \times 5$
10. $6 \times 1\frac{3}{8}$
11. $\frac{5}{7} \times 2\frac{1}{3}$
12. $1\frac{2}{5} \times 1\frac{1}{4}$
13. $2\frac{1}{2} \times 5\frac{4}{5}$
14. $7\frac{2}{3} \times \frac{3}{4}$
15. $2 \times 3\frac{1}{4}$
16. $3\frac{1}{8} \times 1\frac{3}{5}$

3.11 Dividing Fractions

Example 14: $1\frac{3}{4} \div 2\frac{5}{8}$

Step 1: Change the mixed numbers in the problem to improper fractions.

$$1\frac{3}{4} = \frac{(4 \times 1) + 3}{4} = \frac{7}{4} \text{ and } 2\frac{5}{8} = \frac{(8 \times 2) + 5}{8} = \frac{21}{8}.$$

The problem is now $\frac{7}{4} \div \frac{21}{8}$.

Step 2: Invert (turn upside down) the second fraction and multiply. $\frac{7}{4} \times \frac{8}{21}$

Step 3: Cancel where possible and multiply.

$$\frac{\overset{1}{\cancel{7}}}{\underset{1}{\cancel{4}}} \times \frac{\overset{2}{\cancel{8}}}{\underset{3}{\cancel{21}}} = \frac{2}{3}$$

Divide and reduce answers to lowest terms.

1. $2\frac{2}{3} \div 1\frac{7}{9}$
2. $5 \div 1\frac{1}{2}$
3. $1\frac{5}{8} \div 2\frac{1}{4}$
4. $8\frac{2}{3} \div 2\frac{1}{6}$
5. $2\frac{4}{5} \div 2\frac{1}{5}$
6. $3\frac{2}{3} \div 1\frac{1}{6}$
7. $10 \div \frac{4}{5}$
8. $6\frac{1}{4} \div 1\frac{1}{2}$
9. $\frac{2}{5} \div 2$
10. $4\frac{1}{6} \div 1\frac{2}{3}$
11. $9 \div 3\frac{1}{4}$
12. $5\frac{1}{3} \div 2\frac{2}{5}$
13. $4\frac{1}{5} \div \frac{9}{10}$
14. $2\frac{2}{3} \div 4\frac{4}{5}$
15. $3\frac{3}{8} \div 3\frac{6}{7}$
16. $5\frac{1}{4} \div \frac{3}{4}$

3.12 Estimating Multiplying Fractions

Example 15: Estimate $\frac{1}{4}$ of 779.

Step 1: Think of a number close to 779 that can be mentally divided by 4. 779 is close to 800.

Step 2: $\frac{1}{4}$ of $800 = \frac{1}{4} \times \frac{800}{1} = \frac{800}{4} = 800 \div 4 = 200$

$\frac{1}{4}$ of 779 is about 200.

Estimate each of the following.

1. $\frac{1}{2}$ of 589
2. $\frac{1}{4}$ of 391
3. $\frac{1}{3}$ of 877
4. $\frac{1}{5}$ of 488
5. $\frac{1}{6}$ of 1,749
6. $\frac{1}{7}$ of 4,888
7. $\frac{1}{8}$ of 5,555
8. $\frac{1}{5}$ of 3,992
9. $\frac{1}{4}$ of 1,977

3.13 Estimating Reducing Fractions

Example 16: Estimate $\frac{48}{492}$.

Step 1: Find numbers close to 48 and 492 that will make the fraction easy to reduce.
48 is close to 50
492 is close to 500

Step 2: Reduce $\frac{50}{500} = \frac{1}{10}$
The sign $\approx$ means approximately or is about.
$\frac{48}{492} \approx$ (is about) $\frac{1}{10}$

Estimate to reduce the following fractions. The first two are done for you.

1. $\frac{214}{512} \approx \frac{200}{500} = \frac{2}{5}$
2. $\frac{8}{65} \approx \frac{8}{64} = \frac{1}{8}$
3. $\frac{7}{50}$
4. $\frac{5}{46}$
5. $\frac{8}{78}$
6. $\frac{343}{69}$
7. $\frac{21}{79}$
8. $\frac{403}{1592}$
9. $\frac{9}{811}$

3.14 Comparing Fractions

Comparing two fractions means finding which fraction is larger and which fraction is smaller than the other.

Example 17: Compare $\frac{3}{4}$ and $\frac{5}{8}$. Use $>$ or $<$ to show their relationship.

Step 1: Write the two fractions. $\frac{3}{4}$ $\frac{5}{8}$

Step 2: Multiply from the upper left to lower right and write down your answer.

$\frac{3}{4} \searrow \frac{5}{8}$ $3 \times 8 = 24$

Step 3: Multiply from the lower left to upper right and write down your answer after the first answer.

$\frac{3}{4} \nearrow \frac{5}{8}$ 24 20

Step 4: Determine which sign, $>$ or $<$, would fit between 24 and 20.

$24 > 20$

This is the sign to use between the fractions.

$\frac{3}{4} > \frac{5}{8}$ $\frac{3}{4}$ is greater than $\frac{5}{8}$

Fill in the box with the correct sign (>, <, or =).

1. $\frac{7}{9} \square \frac{7}{8}$
2. $\frac{6}{7} \square \frac{5}{6}$
3. $\frac{4}{6} \square \frac{5}{7}$
4. $\frac{3}{10} \square \frac{4}{13}$
5. $\frac{5}{8} \square \frac{4}{11}$
6. $\frac{5}{8} \square \frac{4}{7}$
7. $\frac{9}{10} \square \frac{8}{13}$
8. $\frac{2}{13} \square \frac{1}{10}$
9. $\frac{4}{9} \square \frac{3}{5}$
10. $\frac{2}{6} \square \frac{4}{5}$
11. $\frac{7}{12} \square \frac{6}{11}$
12. $\frac{3}{11} \square \frac{5}{12}$

3.15 Changing Fractions to Decimals

Example 18: Change $\frac{1}{8}$ to a decimal.

Step 1: To change a fraction to a decimal, simply divide the top number by the bottom number.

$$8 \overline{)1}$$

Step 2: Add a decimal point and a 0 after the 1 and divide.

$$\begin{array}{r} 0.1 \\ 8 \overline{)1.0} \\ -8 \\ \hline 2 \end{array}$$

Step 3: Continue adding 0's and dividing until there is no remainder.

$$\begin{array}{r} 0.125 \\ 8 \overline{)1.000} \\ -8 \\ \hline 20 \\ -16 \\ \hline 40 \\ -40 \\ \hline 0 \end{array}$$

In some problems the number after the decimal point begins to repeat. Take, for example, the fraction $\frac{4}{11}$. $4 \div 11 = 0.363636$, and the 36 keeps repeating forever. To show that 36 repeats, simply write a bar above the numbers that repeat, $0.\overline{36}$.

Change the following fractions to decimals.

1. $\frac{4}{5}$
2. $\frac{2}{3}$
3. $\frac{1}{2}$
4. $\frac{5}{9}$
5. $\frac{1}{10}$
6. $\frac{5}{8}$
7. $\frac{5}{6}$
8. $\frac{1}{6}$
9. $\frac{3}{5}$
10. $\frac{7}{10}$
11. $\frac{4}{11}$
12. $\frac{1}{9}$
13. $\frac{7}{9}$
14. $\frac{9}{10}$
15. $\frac{1}{4}$
16. $\frac{3}{8}$
17. $\frac{3}{16}$
18. $\frac{3}{4}$
19. $\frac{8}{9}$
20. $\frac{5}{12}$

3.16 Changing Mixed Numbers to Decimals

If there is a whole number with a fraction, write the whole number to the left of the decimal point. Then change the fraction to a decimal.

Examples: $4\frac{1}{10} = 4.1$ $16\frac{2}{3} = 16.\overline{6}$ $12\frac{7}{8} = 12.875$

Change the following mixed numbers to decimals.

1. $5\frac{2}{3}$	5. $30\frac{1}{3}$	9. $6\frac{4}{5}$	13. $7\frac{1}{4}$	17. $10\frac{1}{10}$
2. $8\frac{5}{11}$	6. $3\frac{1}{2}$	10. $13\frac{1}{2}$	14. $12\frac{1}{3}$	18. $20\frac{2}{5}$
3. $15\frac{3}{5}$	7. $1\frac{7}{8}$	11. $12\frac{4}{5}$	15. $1\frac{5}{8}$	19. $4\frac{9}{10}$
4. $13\frac{2}{3}$	8. $4\frac{9}{100}$	12. $11\frac{5}{8}$	16. $2\frac{3}{4}$	20. $5\frac{4}{11}$

3.17 Changing Decimals to Fractions

Example 19: Change 0.25 to a fraction.

Step 1: Copy the decimal without the point. This will be the top number of the fraction. $\frac{25}{}$

Step 2: The bottom number is a 1 with as many 0's after it as there are digits in the top number. $\frac{25 \leftarrow \text{Two digits}}{100 \leftarrow \text{Two 0's}}$

Step 3: You then need to reduce the fraction. $\frac{25}{100} = \frac{1}{4}$

Examples: $0.2 = \frac{2}{10} = \frac{1}{5}$ $0.65 = \frac{65}{100} = \frac{13}{20}$ $0.125 = \frac{125}{1000} = \frac{1}{8}$

Change the following decimals to fractions.

1. 0.55	5. 0.75	9. 0.71	13. 0.35
2. 0.6	6. 0.82	10. 0.42	14. 0.96
3. 0.12	7. 0.3	11. 0.56	15. 0.125
4. 0.9	8. 0.42	12. 0.24	16. 0.375

3.18 Changing Decimals with Whole Numbers to Mixed Numbers

Example 20: Change 14.28 to a mixed number.

Step 1: Copy the portion of the number that is whole. 14

Step 2: Change .28 to a fraction. $14\frac{28}{100}$

Step 3: Reduce the fraction. $14\frac{28}{100} = 14\frac{7}{25}$

Change the following decimals to mixed numbers.

1. 7.125	5. 16.95	9. 6.7	13. 13.9
2. 99.5	6. 3.625	10. 45.425	14. 32.65
3. 2.13	7. 4.42	11. 15.8	15. 17.25
4. 5.1	8. 15.84	12. 8.16	16. 9.82

3.19 Fraction Word Problems

Solve and reduce answers to lowest terms.

1. Sara buys candy by the pound during the summer. During the first week of summer she buys $1\frac{1}{3}$ pounds of candy, during the second she buys $\frac{3}{4}$ of a pound, and during the third she buys $\frac{4}{5}$ pound. How many pounds did she buy during the first three weeks of summer?
2. Beth has a bread machine that makes a loaf of bread that weighs $1\frac{1}{2}$ pounds. If she makes a loaf of bread for each of her three sisters, how many pounds of bread will she make?
3. Rick chews on a piece of bubble gum for 120 minutes. About every $1\frac{1}{4}$ minutes, he blows a bubble. How many bubbles did Rick make?
4. Juan was competing in a 1,000 meter race, but he had to pull out of the race after running $\frac{3}{4}$ of it. How many meters did he run?
5. Tad needs to measure where the free throw line should be in front of his basketball goal. He knows his feet are $1\frac{1}{8}$ feet long and the free-throw line should be 15 feet from the backboard. How many toe-to-toe steps does Tad need to take to mark off 15 feet?
6. Mary gives her puppy a bath and uses $5\frac{1}{2}$ gallons of water. She throws away $3\frac{2}{3}$ gallons of the water. How much water does she have left?

Chapter 3 Review

Simplify.

1. $\frac{15}{6}$ 2. $\frac{24}{5}$ 3. $\frac{20}{15}$ 4. $\frac{14}{3}$

Reduce.

5. $\frac{9}{27}$ 6. $\frac{4}{16}$ 7. $\frac{8}{12}$ 8. $\frac{12}{18}$

Change to an improper fraction.

9. $5\frac{1}{10}$ 10. 7 11. $3\frac{3}{5}$ 12. $6\frac{2}{3}$

Add and simplify.

13. $\frac{5}{9} + \frac{7}{9}$ 14. $7\frac{1}{2} + 3\frac{3}{8}$ 15. $4\frac{4}{15} + \frac{1}{5}$ 16. $\frac{1}{7} + \frac{3}{7}$

Subtract and simplify.

17. $10 - 5\frac{1}{8}$ 18. $3\frac{1}{3} - \frac{3}{4}$ 19. $9\frac{3}{4} - 2\frac{3}{8}$ 20. $6\frac{1}{5} - 1\frac{3}{10}$

Multiply and simplify.

21. $1\frac{1}{3} \times 3\frac{1}{2}$ 22. $5\frac{3}{7} \times \frac{7}{8}$ 23. $4\frac{4}{6} \times 1\frac{5}{7}$ 24. $\frac{2}{3} \times \frac{5}{6}$

Divide and simplify.

25. $\frac{1}{2} \div \frac{4}{5}$ 26. $6\frac{6}{7} \div 2\frac{2}{3}$ 27. $3\frac{5}{6} \div 11\frac{1}{2}$ 28. $1\frac{1}{3} \div 3\frac{1}{5}$

Estimate the following multiplication and division problems.

29. $\frac{1}{5}$ of 211 30. $\frac{42}{79}$

Find the missing numerators.

31. $\frac{4}{11} = \frac{}{55}$

33. $\frac{2}{7} = \frac{}{35}$

35. $\frac{3}{5} = \frac{}{45}$

37. $\frac{4}{9} = \frac{}{54}$

32. $\frac{3}{8} = \frac{}{48}$

34. $\frac{7}{10} = \frac{}{80}$

36. $\frac{5}{12} = \frac{}{48}$

38. $\frac{3}{11} = \frac{}{121}$

Change to a fraction.

39. 0.55

40. 0.84

41. 0.32

Change to a mixed number.

42. 7.375

43. 9.6

44. 13.25

Change to a decimal.

45. $5\frac{3}{25}$

46. $\frac{7}{100}$

47. $10\frac{2}{3}$

Use the >, <, and = signs to make the following correct.

48. $\frac{5}{6}$ □ $\frac{4}{5}$

49. $\frac{3}{7}$ □ $\frac{4}{8}$

50. $\frac{4}{15}$ □ $\frac{5}{16}$

51. $\frac{3}{4}$ □ $\frac{13}{16}$

52. The Vargas family is hiking a $23\frac{1}{3}$ mile trail. The first day, they hiked $10\frac{1}{2}$ miles. How much further do they have to go to complete the trail?

53. Jena walks $\frac{1}{5}$ of a mile to a friend's house, $1\frac{1}{3}$ miles to the store, and $\frac{3}{4}$ of a mile back home. How far does Jena walk?

54. Cory uses $2\frac{4}{5}$ gallons of paint to mark one mile of this year's spring road race. How many gallons will he use to mark the entire $6\frac{1}{4}$ mile course?

Chapter 3 Test

1. Simplify the improper fraction.

$\frac{27}{5}$

A. $5\frac{4}{5}$

B. $5\frac{1}{5}$

C. $5\frac{2}{5}$

D. $5\frac{3}{5}$

2. Convert $6\frac{2}{4}$ to an improper fraction.

A. $6\frac{1}{2}$

B. $\frac{26}{4}$

C. $\frac{12}{4}$

D. $\frac{16}{4}$

3. Add: $\frac{2}{4} + \frac{5}{8} =$

A. $\frac{7}{12}$

B. $\frac{7}{8}$

C. $1\frac{1}{8}$

D. $1\frac{2}{8}$

4. Subtract: $7\frac{1}{3} - 5 =$

A. $2\frac{1}{3}$

B. $3\frac{2}{3}$

C. $3\frac{1}{3}$

D. $2\frac{2}{3}$

5. Subtract: $12\frac{1}{2} - 6\frac{2}{3} =$

A. $5\frac{1}{6}$

B. $5\frac{2}{6}$

C. $6\frac{1}{3}$

D. $5\frac{5}{6}$

6. Find the product.

$\frac{6}{8} \times \frac{16}{4}$

A. $\frac{22}{12}$

B. 3

C. $\frac{96}{28}$

D. 4

7. Find the quotient.

$\frac{2}{10} \div \frac{8}{5}$

A. $\frac{16}{50}$

B. $\frac{7}{18}$

C. $\frac{1}{8}$

D. $\frac{3}{8}$

8. Which fraction below is greater than $\frac{2}{3}$?

A. $\frac{5}{8}$

B. $\frac{1}{2}$

C. $\frac{7}{9}$

D. $\frac{6}{10}$

9. Melissa's books weigh 14.5 lbs., Jonathan's weigh 12.5 lbs., and Candy's books weigh 7.25 lbs. What is the combined weight of all three books?

A. $33\frac{1}{2}$ lbs.

B. $33\frac{2}{3}$ lbs.

C. $34\frac{1}{4}$ lbs.

D. 34 lbs.

10. Ayesha's dog weighs $40\frac{1}{4}$ lbs. He weighed $8\frac{1}{2}$ lbs. as a puppy. How much weight did her dog gain?

A. 32 lbs.

B. $31\frac{1}{2}$ lbs.

C. $31\frac{3}{4}$ lbs.

D. $30\frac{1}{4}$ lbs.

11. What is the value of $\frac{5}{16}$ in decimal form?

A. 0.3125

B. 0.3025

C. 3.2

D. 3.125

12. What is the value of $12\frac{2}{3}$ in decimal form?

A. 12.25

B. 12.50

C. 12.66

D. 12.75

13. What is the value of 0.78 as a fraction?

A. $\frac{39}{100}$

B. $\frac{7.8}{100}$

C. 78

D. $\frac{39}{50}$

14. What is the value of 1.875 in fractional form?

A. $\frac{7}{5}$

B. $\frac{9}{8}$

C. $\frac{15}{8}$

D. $\frac{7}{6}$

15. Estimate the product. $47 \times 21 =$

A. 800

B. 850

C. 1,000

D. 950

16. Estimate the quotient. $2,364 \div 8 =$

A. 250

B. 300

C. 315

D. 350

Chapter 4
Percents

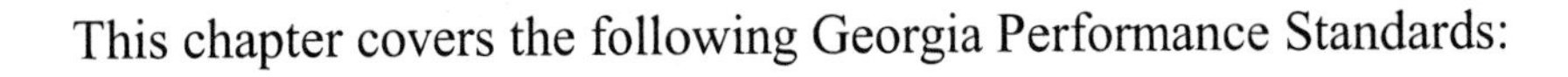
This chapter covers the following Georgia Performance Standards:

M5N	Numbers and Operations	M5N5.a, b
M5P	Process Skills	M5P4.a, b, c M5P5.a, b

4.1 Changing Percents to Decimals and Decimals to Percents

To change a **percent** to a **decimal**, move the **decimal** point two places to the left, and drop the **percent** sign. If there is no decimal point shown, it is understood to be after the number and before the percent sign. Sometimes you will need to add a "0". (See 5% below.)

Example 1: $14\% = 0.14$ $5\% = 0.05$ $100\% = 1$ $103\% = 1.03$

↑ **(decimal point)**

Change the following percents to decimal numbers.

1. 18%	4. 63%	7. 2%	10. 55%	13. 66%	16. 25%	19. 50%
2. 23%	5. 4%	8. 119%	11. 80%	14. 13%	17. 410%	20. 99%
3. 9%	6. 45%	9. 7%	12. 17%	15. 5%	18. 1%	21. 107%

To change a **decimal** to a **percent**, move the decimal two places to the right, and add a percent sign. You may need to add a "0". (See 0.8 below.)

Example 2: $0.62 = 62\%$ $0.07 = 7\%$ $0.8 = 80\%$ $0.166 = 16.6\%$ $1.54 = 154\%$

Change the following decimal numbers to percents.

22. 0.15	25. 0.22	28. 0.648	31. 0.86	34. 0.48	37. 0.375	40. 0.3
23. 0.62	26. 0.35	29. 0.044	32. 0.29	35. 3.089	38. 5.09	41. 2.9
24. 1.53	27. 0.375	30. 0.58	33. 0.06	36. 0.042	39. 0.75	42. 0.06

4.2 Changing Percents to Fractions and Fractions to Percents

Example 3: Change 15% to a fraction.

Step 1: Copy the number without the percent sign. 15 is the top number of the fraction.

Step 2: The bottom number of the fraction is 100.

$15\% = \frac{15}{100}$

Step 3: Reduce the fraction. $\frac{15}{100} = \frac{3}{20}$

Change the following percents to fractions and reduce.

1. 50%	5. 52%	9. 18%	13. 16%	17. 99%
2. 13%	6. 63%	10. 3%	14. 1%	18. 30%
3. 22%	7. 75%	11. 25%	15. 79%	19. 15%
4. 95%	8. 91%	12. 5%	16. 40%	20. 84%

Example 4: Change $\frac{7}{8}$ to a percent.

Step 1: Divide 7 by 8. Add as many 0's as necessary.

```
    0.875
8 ) 7.000
  - 64
     60
  -  56
      40
  -   40
       0
```

Step 2: Change the decimal answer, 0.875, to a percent by moving the decimal point 2 places to the right.

$\frac{7}{8} = 0.875 = 87.5\%$

Change the following fractions to percents.

1. $\frac{1}{5}$	4. $\frac{3}{8}$	7. $\frac{1}{10}$	10. $\frac{3}{4}$	13. $\frac{1}{16}$	16. $\frac{3}{4}$
2. $\frac{5}{8}$	5. $\frac{3}{16}$	8. $\frac{4}{5}$	11. $\frac{1}{8}$	14. $\frac{1}{4}$	17. $\frac{2}{5}$
3. $\frac{7}{16}$	6. $\frac{19}{100}$	9. $\frac{15}{16}$	12. $\frac{5}{16}$	15. $\frac{4}{100}$	18. $\frac{16}{25}$

4.3 Changing Percents to Mixed Numbers and Mixed Numbers to Percents

Example 5: Change 218% to a fraction.

Step 1: Copy the number without the percent sign. 218 is the top number of the fraction.

Step 2: The bottom number of any percent to a fraction problem is always 100.

$$218\% = \frac{218}{100}$$

Step 3: Reduce the fraction, and convert to a mixed number. $\frac{218}{100} = \frac{109}{50} = 2\frac{9}{50}$

Change the following percents to mixed numbers.

1. 150%	6. 163%	11. 205%	16. 340%
2. 113%	7. 275%	12. 405%	17. 199%
3. 222%	8. 191%	13. 516%	18. 300%
4. 395%	9. 108%	14. 161%	19. 125%
5. 252%	10. 453%	15. 179%	20. 384%

Example 6: Change $5\frac{3}{8}$ to a percent.

Step 1: Divide 3 by 8. Add as many 0's as necessary.

$$\begin{array}{r} 0.375 \\ 8\overline{)\,3.000} \\ -\ 2\ 4 \\ \hline 60 \\ -56 \\ \hline 40 \\ -\ 40 \\ \hline 0 \end{array}$$

Step 2: So, $5\frac{3}{8} = 5.375$. Change the decimal answer to a percent by moving the decimal point 2 places to the right and add a % sign.

$5\frac{3}{8} = 5.375 = 537.5\%$

Change the following mixed and whole numbers to percents.

1. $5\frac{1}{2}$	4. $3\frac{1}{4}$	7. $1\frac{3}{10}$	10. $2\frac{13}{25}$	13. $1\frac{3}{16}$	16. $4\frac{4}{5}$
2. $8\frac{3}{4}$	5. $4\frac{7}{8}$	8. $6\frac{1}{5}$	11. $1\frac{1}{8}$	14. $1\frac{1}{16}$	17. $3\frac{2}{5}$
3. 1	6. 3	9. 4	12. 2	15. 5	18. 6

4.4 Comparing the Relative Magnitude of Numbers

When comparing the relative magnitude of numbers, the greater than ($>$), less than ($<$), and the equal to ($=$) signs are the ones most frequently used. The simplest way to compare numbers that are in different notations, like percent, decimals, and fractions, is to change all of them to one notation. Decimals are the easiest to compare.

Example 7: Which is larger: $1\frac{1}{4}$ or 1.3?

Step 1: Change $1\frac{1}{4}$ to a decimal. $\frac{1}{4} = 0.25$, so $1\frac{1}{4} = 1.25$.

Step 2: Compare the two values in decimal form.
$1.25 < 1.3$, so 1.3 is the larger of the two values.

Example 8: Which is smaller: 60% or $\frac{2}{3}$?

Step 1: Change both values to decimals.
$60\% = 0.6$ and $\frac{2}{3} = 0.\overline{66}$

Step 2: Compare the two values in decimal form.
0.6 is smaller than $0.\overline{66}$, so $60\% < \frac{2}{3}$.

Fill in each box with the correct sign.

1. $23.4 \square 23\frac{1}{2}$
2. $17\% \square .17$
3. $\frac{3}{8} \square 37.5\%$
4. $25\% \square \frac{2}{10}$
5. $234\% \square 23.4$
6. $\frac{1}{7} \square 14\%$
7. $13.95 \square 13\frac{8}{9}$
8. $4.0 \square 40\%$
9. $25\% \square \frac{3}{2}$
10. $\frac{12}{4} \square 300\%$
11. $6\% \square \frac{1}{16}$
12. $1.\overline{33} \square \frac{4}{3}$
13. $.8 \square \frac{4}{5}$
14. $75\% \square \frac{3}{4}$
15. $\frac{5}{8} \square 62\%$

Compare the sums, differences, products, and quotients below. Fill in each box with the correct sign.

16. $(32 + 15) \square (65 - 17)$
17. $(45 - 13) \square (31 + 9)$
18. $(24 \div 4) \square (24 \div 6)$
19. $(48 \div 6) \square (4 \times 3)$
20. $(4 \times 3) \square (48 \div 6)$
21. $(18 \times 4) \square (5 \times 17)$
22. $[(1 + 3) + 5] \square [5 + (3 + 1)]$
23. $[1 + (3 + 5)] \square [(5 - 3) + 1]$
24. $(25 \div 5) \square (5 \times 5)$
25. $(6 + 4 + 2) \square [(6 + 4) + 2]$

4.5 Modeling Percentages

Example 9: Find the percent of the area shaded in the following rectangle.

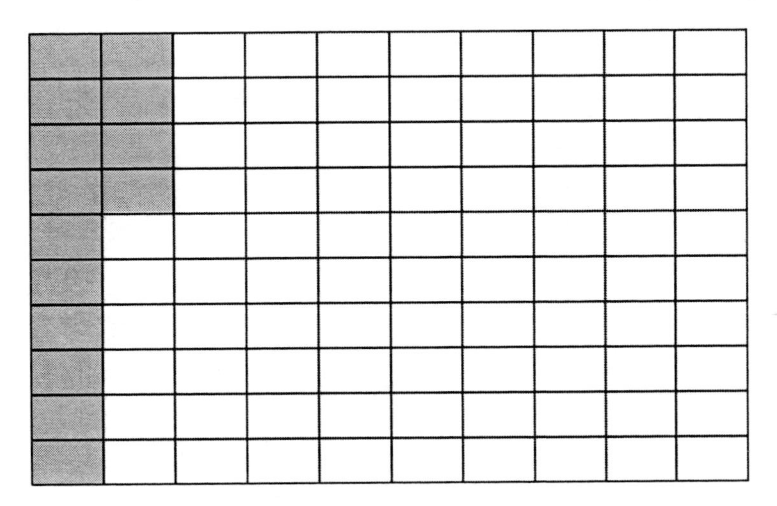

Step 1: Count the number of boxes in the rectangle. There are 100 boxes. That means each shaded box represents 1/100 of 100% or 1%.

Step 2: Count the number of boxes in the rectangle that are shaded. Fourteen boxes are shaded.

Step 3: There are $\frac{14}{100}$ shaded. $\frac{14}{100} = 14\%$

Example 10: Find the percent of the area shaded in the following circle.

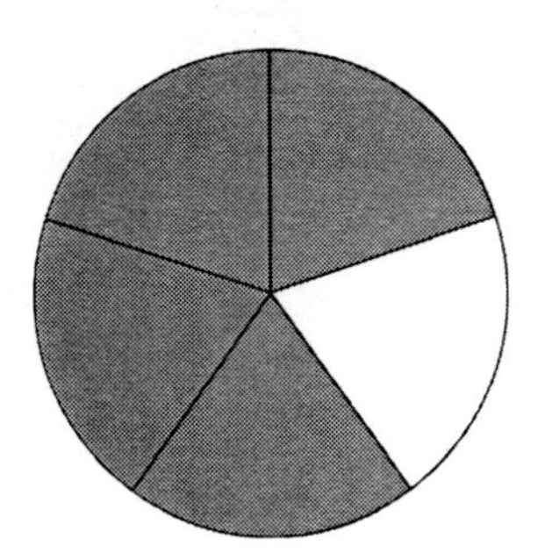

Step 1: Count the number of divisions in the circle. There are five divisions.

Step 2: Count the number of divisions in the circle that are shaded. Four sections are shaded. 4 out of 5 are shaded

Step 3: 4 out of 5 can be represented as $\frac{4}{5} = 0.80 = 80\%$

Find the following percentage of each shaded area.

1. 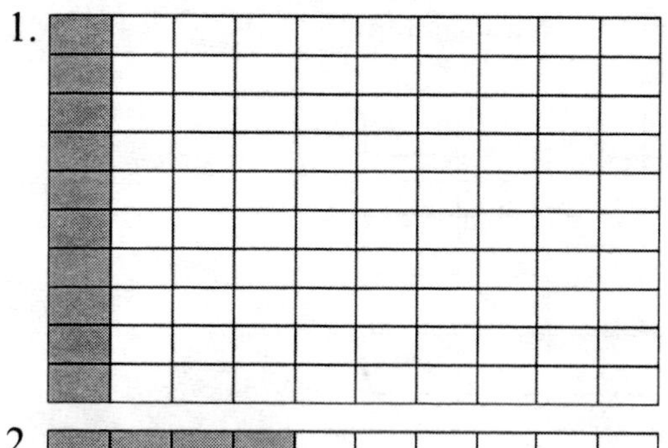

2. 

3. 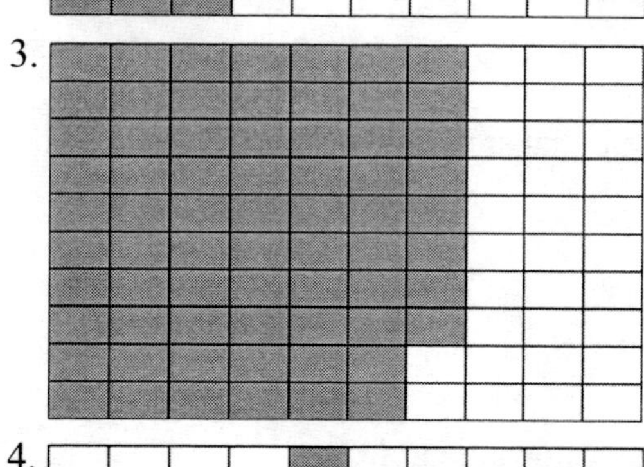

4. 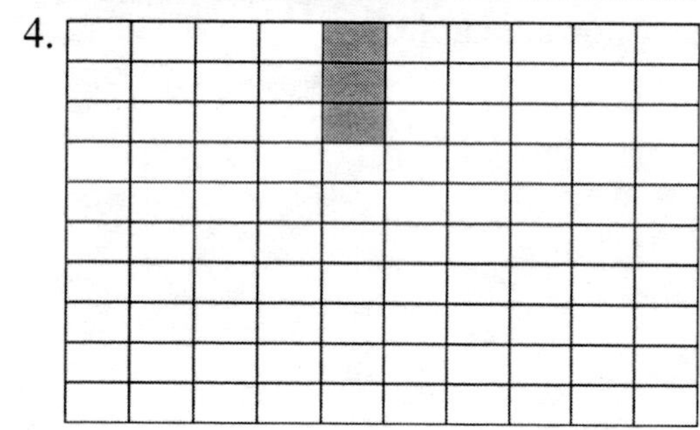

5. 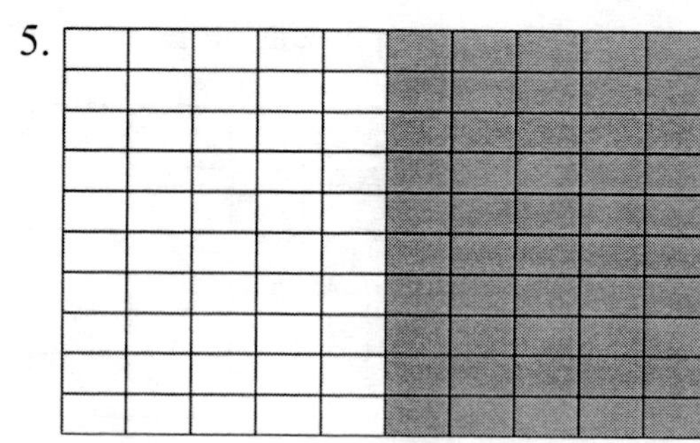

6. 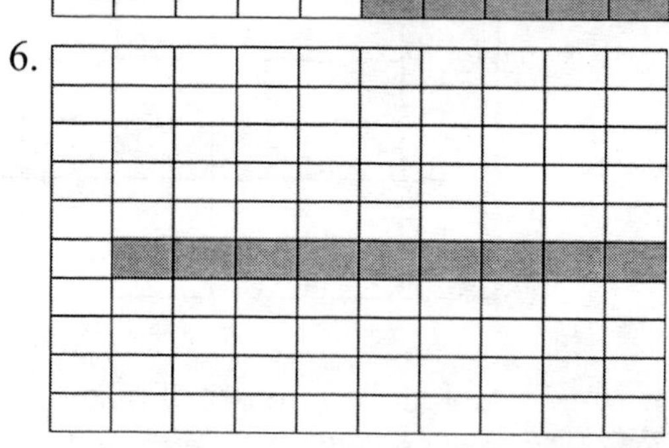

7. 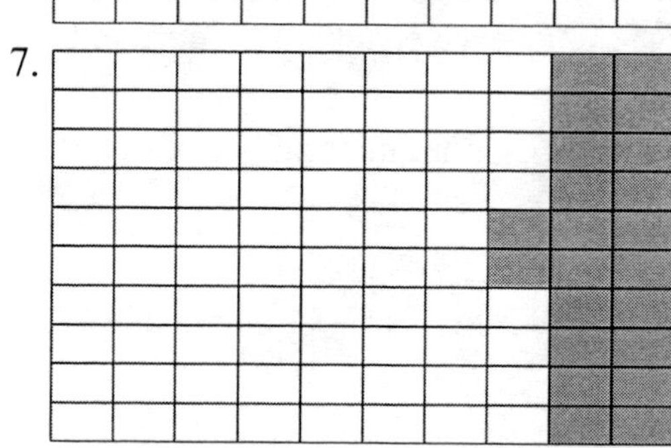

8. 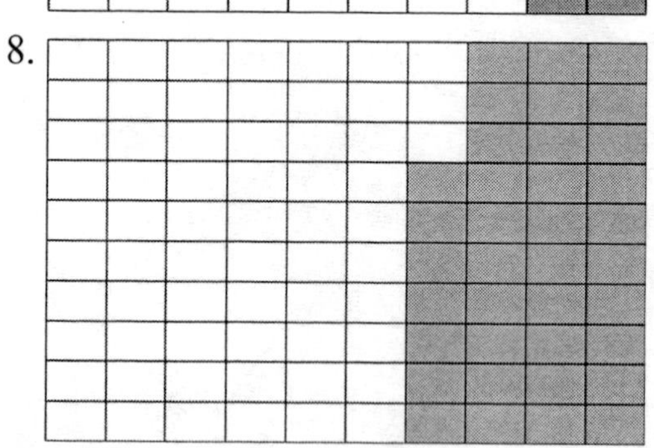

Find the following percentage of each shaded area.

9. 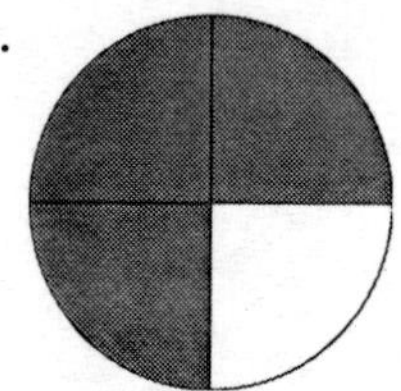

10. 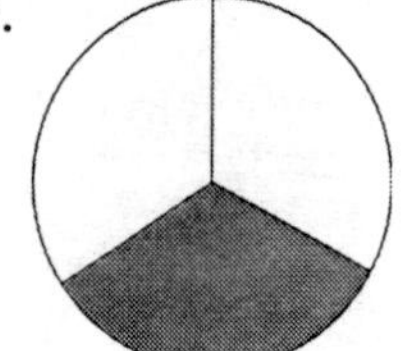

11. 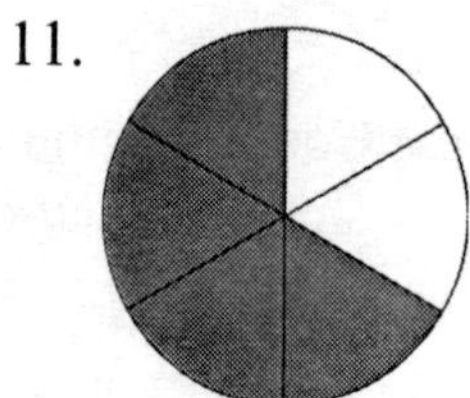

4.6 Representing Rational Numbers Graphically

You now know how to convert fractions to decimals, decimals to fractions, fractions to percentages, percentages to fractions, decimals to percentages, and percentages to decimals. Study the examples below to understand how fractions, decimals, and percentages can be expressed graphically.

Examples:

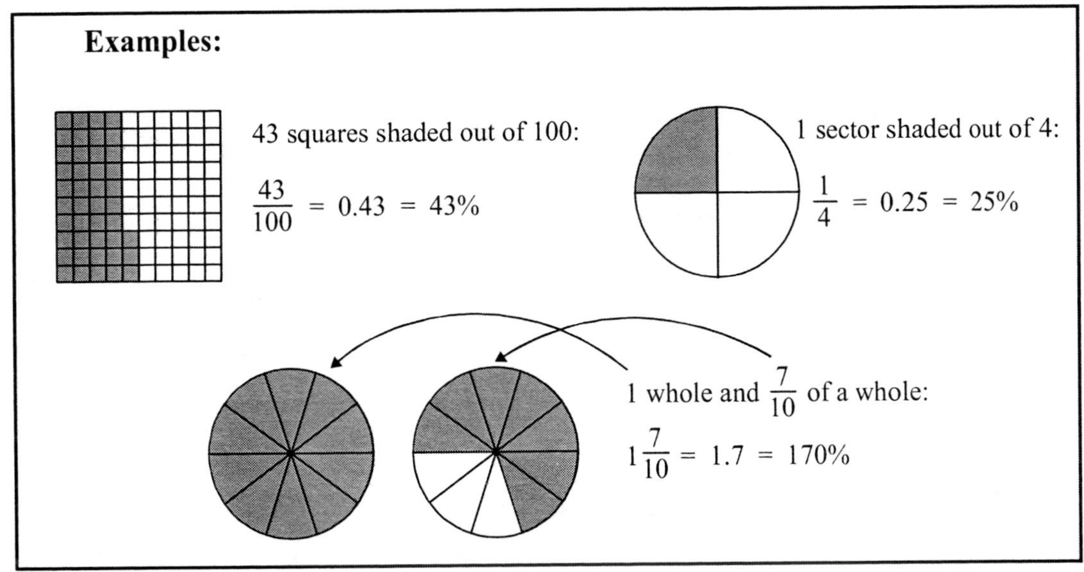

Fill in the missing information in the chart below. Shade in the graphic for the problems that are not shaded for you. Reduce all fractions to lowest terms.

Graphic	Fraction	Decimal	Percent	Graphic	Fraction	Decimal	Percent
1.				5.			125%
2.		0.92		6.			
3.				7.			
4.	$\frac{4}{5}$			8.			

4.7 Percent Word Problems

Example 11: Three out of four students prefer pizza for lunch over hot dogs. What percent of the students prefer pizza?

Step 1: Change three out of four to a fraction. $\frac{3}{4}$

Step 2: Change $\frac{3}{4}$ to a percent. $\frac{3}{4} = 75\%$ of the students prefer pizza.

For each of the following, find the percent.

1. 18 out of 24 students take a music class in middle school.
2. 6 out of the 30 students get a grade of an A in the class.
3. 3 out of 10 students ride the bus to school.
4. 16 out of 20 students in my first period ate breakfast.
5. Out of 12 cats, 3 of them have stripes.
6. In a neighborhood store, 8 out of 50 customers have an infant in their cart.
7. 240 out of 500 people at the fair are over 21.
8. In Alaska, 9 months out of 12 record temperatures below 0° C.
9. He hits the ball 360 times out of 500 times at bat.
10. She scores 552 out of a possible 600 points on her math test.
11. 16 out of 40 pieces in the box of candy are covered in dark chocolate.
12. 680 out of 800 students have not had a cold this year.
13. 7 out of 8 of the flowers are pink.
14. 312 out of 600 men at the football game have on earmuffs.
15. 824 out of 1,000 tulips are red.
16. Kristen sells 475 out of her 500 boxes of Girl Scout cookies.
17. 459 out of 675 students in the 6th grade went to the spring dance this year.
18. In a bag of 60 candy coated chocolate pieces, 9 are yellow.

4.8 Missing Information

Problems can only be solved if you are given enough information. In some cases, you are not given all of the information needed to solve a problem. To figure out what information is missing, you are developing a mathematical argument.

Example 12: Chuck has worked on his job for 1 year now. At the end of a year, his employer gave him a 12% raise. How much does Chuck make now?
To solve this problem, you need to know how much Chuck made when he began his job one year ago.

Find what is missing from each problem.

1. Fourteen percent of the coated chocolate candies in Amin's bag were yellow. At that rate, how many of the candies were yellow?

2. Patrick is putting up a fence around all four sides of his back yard. The fence costs $2.25 per foot and his yard is 150 feet wide. How much will the fence cost?

3. Yoko worked 5 days last week. She made $6.75 per hour before taxes. How much did she make last week before taxes were taken out?

4. Which is a better buy: a 4 oz. bar of soap for $0.88 or a bath bar for $1.20?

5. Randy bought a used car for $4568 plus sales tax. What was the total cost of the car?

6. The Portes family ate at a restaurant, and each of the dinners cost $5.95. They left a 15% tip. What was the total amount of the tip?

7. If a kudzu plant grows three feet per day, in what month will it be 90 feet long?

8. Bethany traveled by car to her sister's house in Valdosta. She traveled at an average speed of 52 miles per hour. She arrived at 4:00 P.M. How far did she travel?

9. Terrence earns $7.50 per hour plus 5% commission on total sales over $500 per day. Today he sold $6, 500 worth of merchandise. How much did he earn for the day?

10. Michelle works at a department store and gets an employee's discount on all her purchases. She wants to buy a sweater that sells for $38. How much will the sweater cost after her discount?

11. Matsu filled his car with 10 gallons of gas and paid for the gas with a $20 bill. How much change did he get back?

12. Olivia budgets $5.00 per work day for lunch. How much does she budget for lunch each month?

4.9 Sales Tax

Example 13: The total price of a sofa is $560.00 × 6% **sales tax**. How much is the sales tax? What is the total cost?

Step 1: You will need to change 6% to a decimal.

$6\% = 0.06$

Step 2: Simply multiply the cost, $560, by the tax rate, 6%. $560 \times 0.06 = 33.6$
The answer will be $33.60. (You need to add a 0 to the answer. When dealing with money, there must be two places after the decimal point.)

$$\begin{array}{rr} & \textbf{COST} \\ \times & \textbf{6\% TAX} \\ \hline & \textbf{SALES TAX} \end{array} \qquad \begin{array}{rr} & \textbf{\$560} \\ \times & \textbf{0.06} \\ \hline & \textbf{\$33.60} \end{array}$$

Step 3: Add the sales tax amount, $33.60, to the cost of the item sold, $560. This is the total cost.

$$\begin{array}{rr} & \textbf{COST} \\ + & \textbf{SALES TAX} \\ \hline & \textbf{TOTAL COST} \end{array} \qquad \begin{array}{rr} & \textbf{\$560.00} \\ + & \textbf{33.60} \\ \hline & \textbf{\$593.60} \end{array}$$

Note: When the answer to the question involves money, you always need to round off the answer to the nearest hundredth (2 places after the decimal point). Sometimes you will need to add a zero.

Figure the total costs in the problems below. The first one is done for you.

ITEM	PRICE	% TAX	MULTIPLY	PRICE PLUS TAX	TOTAL
1. jeans	$42	7%	$42 × 0.07 = $2.94	42 + 2.94 = 44.94	$44.94
2. truck	$17,495	6%			
3. film	$5.89	8%			
4. T-shirt	$12	5%			
5. football	$36.40	4%			
6. soda	$1.78	5%			
7. 4 tires	$105.80	10%			
8. clock	$18	6%			
9. burger	$2.34	5%			
10. software	$89.95	8%			

Chapter 4 Review

Change the following percents to decimals.

1. 45%
2. 219%
3. 22%
4. 1.25%

Change the following decimals to percents.

5. 0.52
6. 0.64
7. 1.09
8. 0.625

9. What is 1.65 written as a percent?

10. Change 5.65 to a percent.

Change the following percents to fractions.

11. 25%
12. 3%
13. 68%
14. 102%

Change the following fractions to percents.

15. $\frac{9}{10}$
16. $\frac{5}{16}$
17. $\frac{1}{8}$
18. $\frac{1}{4}$

Find the following percentage of each shaded area.

19.

20.

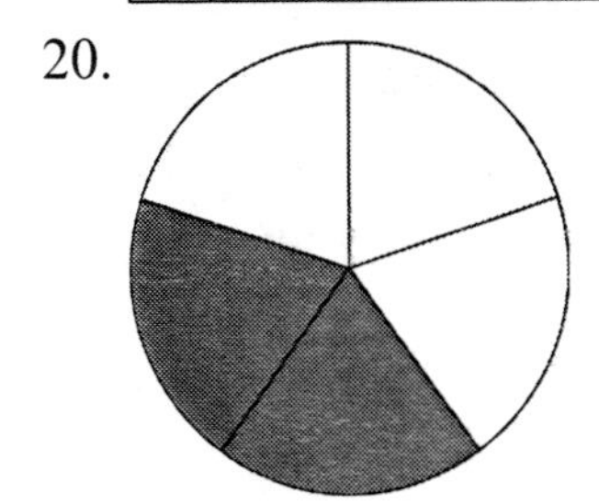

21.

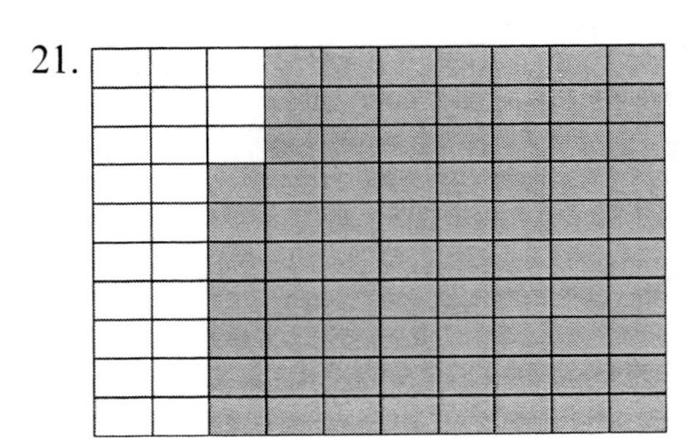

22.

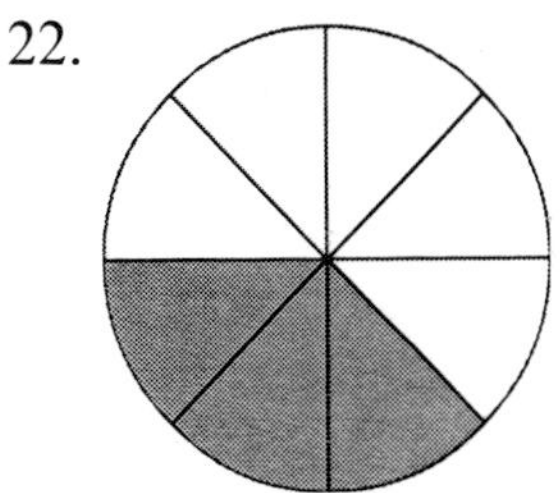

Chapter 4 Test

1. Convert 68% to a decimal.

A. 0.68
B. 6.8
C. 0.068
D. 68

2. Convert 0.046 to a percent.

A. 460%
B. 4%
C. 46%
D. 4.6%

3. Convert $\frac{5}{8}$ to a percent.

A. 40%
B. 62.5%
C. 47%
D. 60%

4. What is the value of 575% as a fraction?

A. $\frac{5}{75}$
B. $\frac{75}{15}$
C. $\frac{23}{4}$
D. $\frac{19}{5}$

5. What is the value of 405% as a mixed number?

A. $4\frac{1}{4}$
B. $4\frac{1}{2}$
C. $4\frac{1}{20}$
D. $4\frac{1}{10}$

6. Convert 112% to a mixed number.

A. $1\frac{3}{25}$
B. $12\frac{1}{100}$
C. $1\frac{5}{8}$
D. $1\frac{3}{4}$

7. Which number is less than 25%?

A. $\frac{2}{3}$
B. 0.33
C. $\frac{1}{5}$
D. 0.5

8. The shaded portion represents what percentage?

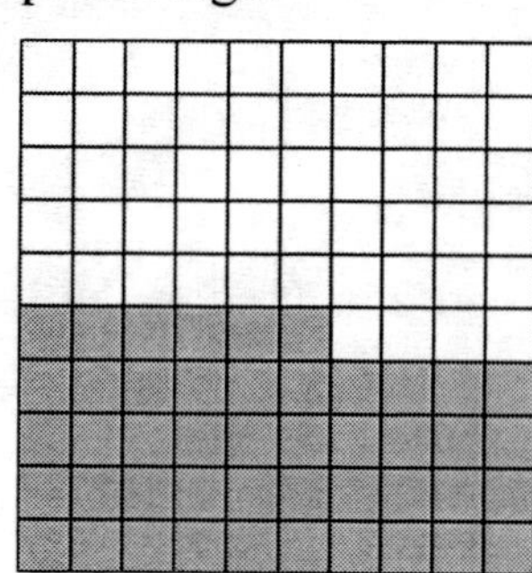

A. 4.6%
B. 46%
C. 0.46%
D. 0.046%

9. James is purchasing a bike for $67.00. If there is a 5% charge for shipping and handling, what is the total cost of the bike?

A. $67.50
B. $70.35
C. $64.35
D. none of the above

10. What missing information is needed to solve this problem?
Carter earns $5.00 per hour working as a cashier. He earned enough this week to purchase an electric scooter for $200. After the purchase, how much money does he have left?

A. total days worked
B. hours worked this week
C. original price of the scooter
D. none of the above

11. Tate is saving his money to buy a go-cart. The go-cart he wants costs $240. His mother agreed to chip in $60. What percent of the total cost of the go-cart is his mother contributing?

A. 20%
B. 25%
C. 40%
D. 75%

12.

Fish Palace

Going Out of Business Sale

All fish, aquariums, & accessories

65% off

According to the ad above, how much would you save by buying an aquarium that regularly sold for $54.00?

A. $18.90
B. $20.00
C. $32.50
D. $35.10

13. Out of the 144 players participating in Little League baseball, 18 were picked for the All-Star game. What percent of the total were picked for the All-Star game?

A. 8%
B. 12.5%
C. 14.4%
D. 18%

14. Renata bought a stuffed bear priced at $14.80 and paid 5% sales tax. What was the total price of her purchase?

A. $14.06
B. $14.85
C. $15.00
D. $15.54

15. The winning pie-eating contestant ate 92% of the 25 pies. How many pies did the contestant eat?

A. 20
B. 21
C. 22
D. 23

Chapter 5
Introduction to Algebra

This chapter covers the following Georgia Performance Standards:

M5A	Algebra	M5A1.a, b, c
M5P	Process Skills	M5P3.a, d

5.1 Algebra Vocabulary

Vocabulary Word	Example	Definition
variable	$4x$ (x is the variable)	a letter that can be replaced by a number
coefficient	$4x$ (4 is the coefficient)	a number multiplied by a variable or variables
term	$5x^2 + x - 2$ ($5x^2$, x, and -2 are terms)	numbers or variables separated by + or − signs
constant	$5x + 2y + 4$ (4 is a constant)	a term that does not have a variable
degree	$4x^2 + 3x - 2$ (the degree is 2)	the largest power of a variable in an expression
leading coefficient	$4x^2 + 3x - 2$ (4 is the leading coefficient)	the number multiplied by the term with the highest power
sentence	$2x = 7$ or $5 \leq x$	two algebraic expressions connected by $=, \neq, <, >, \leq, \geq$, or $\approx$
equation	$4x = 8$	a sentence with an equal sign
inequality	$7x < 30$ or $x \neq 6$	a sentence with one of the following signs: $\neq, <, >, \leq, \geq$, or $\approx$
base	6^3 (6 is the base)	the number used as a factor
exponent	6^3 (3 is the exponent)	the number of times the base is multiplied by itself

5.2 Order of Operations

In long math problems with $+$, $-$, $\times$, $\div$, (), and exponents in them, you have to know what to do first. Without following the same rules, you could get different answers. If you will memorize the silly sentence, Please Excuse My Dear Aunt Sally, you can memorize the order you must follow.

Please "**P**" stands for parentheses. You must get rid of parentheses first.
Examples: $3(1+4) = 3(5) = 15$
$6(10-6) = 6(4) = 24$

Excuse "**E**" stands for exponents. You must eliminate exponents next.
Example: $4^2 = 4 \times 4 = 16$

My Dear "**M**" stands for multiply. "**D**" stands for divide. Start on the left of the equation and perform all multiplications and divisions in the order in which they appear.

Aunt Sally "**A**" stands for add. "**S**" stands for subtract. Start on the left and perform all additions and subtractions in the order they appear.

Example 1: $12 \div 2(6-3) + 3 - 1$

Please	Eliminate **parentheses**. $6 - 3 = 3$ so now we have	$12 \div 2 \times 3 + 3 - 1$
Excuse	Eliminate **exponents**. We do not have any exponents.	$12 \div 2 \times 3 + 3 - 1$
My Dear	**Multiply** and **divide** next in order from left to right.	$12 \div 2 = 6$ then $6 \times 3 = 18$
Aunt Sally	Last, we **add** and **subtract** in order from left to right.	$18 + 3 - 1 = 20$

Simplify the following problems.

1. $6 + 9 \times 2 - 4$
2. $3(4+2) - 6$
3. $3(6-3) - 2$
4. $49 \div 7 - 3 \times 2$
5. $10 \times 4 - (7-2)$
6. $2 \times 3 \div 6 \times 4$
7. $4 \div 8(4+2)$
8. $7 + 8(14-6) \div 4$
9. $(2+8-8) \times 4$
10. $4(8-5) \times 4$
11. $8 + 4 \times 2 - 6$
12. $3(4+6) + 3$
13. $(12-6) + 27 \div 3$
14. $82 - 1 + 4 \div 2$
15. $1 - (2-1) + 8$
16. $12 - 2(7-2)$
17. $18 \div (6+3) - 1$
18. $10 + 3 - 2 \times 3$
19. $4 + (7+2) \div 3$
20. $7 \times 4 - 9 \div 3$

5.3 Substituting Numbers for Variables

These problems may look difficult at first glance, but they are very easy. Simply replace the variable with the number the variable is equal to, and solve the problems.

Example 2: In the following problems, substitute 10 for a.

Problem	Calculation	Solution
1. $a+1$	Simply replace the a with 10. $10+1$	11
2. $17-a$	$17-10$	7
3. $9a$	This means multiply. 9×10	90
4. $\frac{30}{a}$	This means divide. $30 \div 10$	3
5. a^3	$10 \times 10 \times 10$	1000
6. $5a+6$	$(5 \times 10)+6$	56

Note: Be sure to do all multiplying and dividing before adding and subtracting.

Example 3: In the following problems, let $x=2$, $y=4$, and $z=5$.

Problem	Calculation	Solution
1. $5xy+z$	$5 \times 2 \times 4+5$	45
2. $xz+5$	$2 \times 5+5=10+5$	15
3. $\frac{yz}{x}$	$(4 \times 5) \div 2=20 \div 2$	10

In the following problems, $t=7$. Solve the problems.

1. $t+3=$
2. $18-t=$
3. $\frac{21}{t}=$
4. $3t-5=$
5. $t+1=$
6. $2t-4=$
7. $9t \div 3=$
8. $\frac{t}{7}=$
9. $5t+6=$
10. $\frac{(t-7)}{6}=$
11. $4t+5t=$
12. $\frac{6t}{3}=$

In the following problems $a=4$, $b=2$, $c=5$, and $d=10$. Solve the problems.

13. $4a+2c=$
14. $3bc-d=$
15. $\frac{ac}{d}=$
16. $d-2a=$
17. $a-b=$
18. $abd=$
19. $5c-a=$
20. $cd+bc=$
21. $\frac{6b}{a}=$
22. $9a+b=$
23. $5+3bc=$
24. $d+d+1=$

5.4 Understanding Algebra Word Problems

The biggest challenge to solving word problems is figuring out whether to add, subtract, multiply, or divide. Below is a list of key words and their meanings. This list does not include every situation you might see, but it includes the most common examples.

Words Indicating Addition	Example	Add
and	6 **and** 8	$6 + 8$
increased	The original price of $15 **increased** by $5.	$15 + 5$
more	3 coins and 8 **more**	$3 + 8$
more than	Josh has 10 points. Will has 5 **more than** Josh.	$10 + 5$
plus	8 baseballs **plus** 4 baseballs	$8 + 4$
sum	the **sum** of 3 and 5	$3 + 5$
total	the **total** of 10, 14, and 15	$10 + 14 + 15$

Words Indicating Subtraction	Example	Subtract
decreased	$16 **decreased** by $5	$16 - 5$
difference	the **difference** between 18 and 6	$18 - 6$
less	14 days **less** 5	$14 - 5$
less than	Jose completed 2 laps **less than** Mike's 9.	*$9 - 2$
left	Ray sold 15 out of 35 tickets. How many did he have **left**?	*$35 - 15$
lower than	This month's rainfall is 2 inches **lower than** last month's rainfall of 8 inches.	*$8 - 2$
minus	15 **minus** 6	$15 - 6$

* In subtraction word problems, you cannot always subtract the numbers in the order that they appear in the problem. Sometimes the first number should be subtracted from the last. You must read each problem carefully.

Words Indicating Multiplication	Example	Multiply
double	Her $1,000 profit **doubled** in in a month.	1000×2
half	**Half** of the $600 collected went to charity.	$\frac{1}{2} \times 600$
product	the **product** of 4 and 8	4×8
times	Li scored 3 **times** as many points as Ted who only scored 4.	3×4
triple	The bacteria **tripled** its original colony of 10,000 in just one day.	$3 \times 10,000$
twice	Ron has 6 CDs. Tom has **twice** as many.	2×6

Words Indicating Division	Example	Divide
divide into, by, or among	The group of 70 **divided into** 10 teams	$70 \div 10$ or $\frac{70}{10}$
quotient	the **quotient** of 30 and 6	$30 \div 6$ or $\frac{30}{6}$

Match the phrase with the correct algebraic expression below. The answers will be used more than once.

A. $y - 2$

B. $2y$

C. $y + 2$

D. $\frac{y}{2}$

E. $2 - y$

1. 2 more than y
2. 2 divided into y
3. 2 less than y
4. twice y
5. the quotient of y and 2
6. y increased by 2
7. 2 less y
8. the product of 2 and y
9. y decreased by 2
10. y doubled
11. 2 minus y
12. the total of 2 and y

Now practice writing parts of algebraic expressions from the following word problems.

Example 4: the product of 3 and a number, t Answer: $3t$

13. 3 less than x
14. y divided among 10
15. the sum of t and 5
16. n minus 14
17. 5 times k
18. the total of z and 12
19. double the number b
20. x increased by 1
21. the quotient t and 4
22. half of a number, y
23. bacteria culture, b, doubled
24. triple John's age, y
25. a number, n, plus 4
26. quantity, t, less 6
27. 18 divided by a number, x
28. n feet lower than 10
29. 3 more than p
30. the product of 4 and m
31. a number, y, decreased by 20
32. 5 times as much as x

Chapter 5 Review

Solve the following problems using $x = 2$.

1. $3x + 4 =$

2. $\frac{6x}{4} =$

3. $3x - 5 =$

4. $\frac{x + 8}{2} =$

5. $12 - 3x =$

6. $x - 1 =$

7. $5x + 4 =$

8. $9 - x =$

9. $2x + 2 =$

Solve the following problems. Let $w = 1$, $y = 3$, $z = 5$.

10. $5w - y =$

11. $wyz + 2 =$

12. $z - 2w =$

13. $\frac{3z + 5}{wz} =$

14. $\frac{6w}{y} + \frac{z}{w} =$

15. $25 - yz =$

16. $2y + 3$

17. $4w - (yw) =$

18. $7y - 4z =$

Chapter 5 Test

1. Solving the following expression using $a = 4$.

   $3a - 2$

   A. 14
   B. 10
   C. 16
   D. 12

2. Solving the following expression using $y = 3$.

   $8y \div 3$

   A. 8
   B. 21
   C. 24
   D. 72

3. Write the expression from the following word problem.

   A number divided by the sum of nine and two.

   A. $\frac{x}{9+2}$
   B. $\frac{9+2}{x}$
   C. $\frac{x}{9-2}$
   D. $\frac{x+2}{9}$

4. Use order of operations to solve the following expression:

   $8 \times 10 - 7$

   A. 8
   B. 21
   C. 80
   D. 73

5. Solving the following expression using $x = 2$ and $y = 5$.

   $3x + 4y - 1$

   A. 22
   B. 13
   C. 25
   D. 10

6. Write the expression from the following word problem.

   Five less than x plus seven.

   A. $5 - (x + 7)$
   B. $(x + 5) - 7$
   C. $(x + 7) - 5$
   D. $(x - 7) + 5$

7. Write the expression from the following word problem.

   Fifteen minus a number, then divided by two equals eleven.

   A. $\frac{15 - y}{2} = 11$
   B. $11 - \frac{y}{2} = 15$
   C. $15 - \frac{y}{2} = 11$
   D. $2 - \frac{y}{15} = 11$

8. Use order of operations to solve the following expression:

   $(10 + 5) \div 5 - 2$

   A. 9
   B. 5
   C. 1
   D. Cannot be solved.

Chapter 6
Measurement

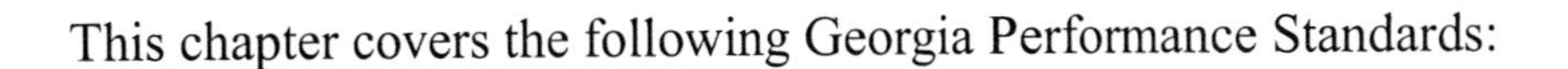
This chapter covers the following Georgia Performance Standards:

M5M	Measurement	M5M3.a, b
M5P	Process Skills	M5P1.a, b

6.1 Customary Measure

Customary measure in the United States is based on the English system. The following chart gives common customary units of measure as well as the standard units for time.

English System of Measure

Measure	Abbreviations	Appropriate Instrument
Time: 1 week = 7 days 1 day = 24 hours 1 hour = 60 minutes 1 minute = 60 seconds	 week = wk hour = hr or h minutes = min seconds = sec	 calendar clock clock clock
Length: 1 mile = 5,280 feet 1 yard = 3 feet 1 foot = 12 inches	 mile = mi yard = yd foot = ft inch = in	 odometer yard stick, tape line ruler, yard stick
Volume: 1 gallon = 4 quarts 1 quart = 2 pints 1 pint = 2 cups 1 cup = 8 ounces	 gallon = gal quart = qt pint = pt ounce = oz	 quart or gallon container quart container cup, pint, or quart container cup
Weight: 1 pound = 16 ounces	 pound = lb ounce = oz	 scale or balance
Temperature: Fahrenheit Celsius	 °F °C	 thermometer thermometer

6.2 Approximate English Measure

Match the item on the left with its approximate (not exact) measure on the right. You may use some answers more than once.

1. The height of an average woman is about _____.
2. An average candy bar weighs about _____.
3. An average doughnut is about _____ across (in diameter).
4. A piece of notebook paper is about _____ long.
5. A tennis ball is about _____ across (in diameter).
6. The average basketball is about _____ across.
8. How long is the average lunch table?
9. About how much does this book weigh?
10. What is the average height of a table?

A. 1 yard
B. 2 yards
C. $5\frac{1}{2}$ feet
D. 4 weeks
E. $2\frac{1}{2}$ inches
F. 2 ounces
G. 1 foot
H. 1 pound

6.3 Converting Units of Measure

English System of Measure:

Time	**Abbreviations**
1 week = 7 days	week = wk
1 day = 24 hours	hour = hr or h
1 hour = 60 minutes	minutes = min
1 minute = 60 seconds	seconds = sec

Length	**Abbreviations**
1 mile = 5,280 feet	mile = mi
1 yard = 3 feet	yard = yd
1 foot = 12 inches	foot = ft
	inches = in

Volume	**Abbreviations**
1 gallon = 4 quarts	gallon = gal
1 quart = 2 pints	quart = qt
1 pint = 2 cups	pint = pt
1 cup = 8 ounces	ounce = oz

Weight	**Abbreviations**
16 ounces = 1 pound	pound = lb
	ounce = oz

Example 1: Simplify: 2 days 34 hr 75 min

Step 1: 75 minutes is more than 1 hour. There are 60 minutes in an hour so divide 75 by 60.

$$60\overline{)75} = 1 \text{ hr}, \quad 75 - 60 = 15 \text{ min}$$

$$\begin{array}{rrrr} & 2 \text{ days} & 34 \text{ hr} & \cancel{75} \text{ min} \\ + & & 1 \text{ hr} & 15 \text{ min} \\ \hline & 2 \text{ days} & 35 \text{ hr} & 15 \text{ min} \end{array}$$

Step 2: 35 hours is more than 1 day. There are 24 hours in a day so divide 35 hours by 24.

$$24\overline{)35} = 1 \text{ day}, \quad 35 - 24 = 11 \text{ hr}$$

$$\begin{array}{rrrr} & 2 \text{ days} & \cancel{35} \text{ hr} & 15 \text{ min} \\ + & 1 \text{ day} & 11 \text{ hr} & \\ \hline & 3 \text{ days} & 11 \text{ hr} & 15 \text{ min} \end{array}$$

2 days 34 hr 75 min = 3 days 11 hr 15 min

Simplify the following.

1. 3 lb 20 oz
2. 2 cup 12 oz
3. 3 wk 9 days 30 hr
4. 1 pt 1 cup 16 oz
5. 2 hr 84 min 62 sec
6. 1 gal 6 qt 3 pt
7. 3 yd 10 ft 18 in
8. 2 wk 8 days 36 hours
9. 2 ft 18 in
10. 1 lb 33 oz
11. 23 hr 62 min 94 sec
12. 3 days 54 hr 75 min

6.4 Converting in the Customary System

Abbreviations:

mile = mi	gallon = gal	pound = lb
yard = yd	quart = qt	ounce = oz
feet = ft	pint = pt	
inch = in	cup = c	

Equivalents:

1 yard = 3 feet = 36 inches	1 gallon = 4 quarts
1 foot = 12 inches	1 quart = 4 cups = 2 pints
	1 pint = 2 cups = 16 ounces
1 pound = 16 ounces	1 cup = 8 ounces

Using the information above, fill in the blanks below.

1. 32 ounces = ________ pound(s)
2. 36 inches = ________ yard(s)
3. 3 yards = ________ feet
4. 2 gallons = ________ quart(s)
5. 5 pints = ________ cup(s)
6. 1 gallon = ________ quart(s)
7. 1 pint = ________ ounce(s)
8. 2 yards = ________ inches
9. 60 inches = ________ feet
10. 2 quarts = ________ gallon(s)
11. 1 quart = ________ ounces
12. 1 quart = ________ cups
13. 3 pints = ________ ounces
14. 18 inches = ________ feet
15. 3 gallons = ________ pint(s)
16. 5 cups = ________ pints
17. 5 quarts = ________ gallon(s)
18. 4 feet = ________ yard(s)
19. 8 pints = ________ gallon(s)
20. 6 cups = ________ quarts
21. 1 pound = ________ ounces
22. 1 yard = ________ inches
23. 7 feet = ________ inches
24. 7 pints = ________ gallons
25. 5 pints = ________ gallons
26. 6 feet = ________ yards
27. 1 gallon = ________ quarts
28. 2 pounds = ________ ounces
29. 6 pints = ________ quarts
30. 3 cups = ________ ounces

6.5 The Metric System

The metric system uses units based on multiples of ten. The basic units of measure in the metric system are the meter, the liter, and the gram. Metric prefixes tell what multiple of ten the basic unit is multiplied by. Below is a chart of metric prefixes and their values. The ones rarely used are shaded.

Prefix	**kilo (k)**	**hecto (h)**	**deka (da)**	**unit (m, L, g)**	**deci (d)**	**centi (c)**	**milli (m)**
Meaning	**1000**	**100**	**10**	**1**	**0.1**	**0.01**	**0.001**

Multiply when changing from a greater unit to a smaller one; **divide** when changing from a smaller unit to a larger one. **The chart is set up to help you know how far and which direction to move a decimal point when making conversions from one unit to another.**

6.6 Understanding Meters

The basic unit of **length** in the metric system is the **meter**. Meter is abbreviated "m".

Metric Unit	Abbreviation	Memory Tip	Equivalents
1 millimeter	mm	Thickness of a dime	10 mm = 1 cm
1 centimeter	cm	Width of the tip of the little finger	100 cm = 1 m
1 meter	m	Distance from the nose to the tip of fingers (a little longer than a yard)	1000 m = 1 km
1 kilometer	km	A little more than half a mile	

6.7 Understanding Liters

The basic unit of **liquid volume** in the metric system is the **liter**. Liter is abbreviated "L".

The liter is the volume of a cube measuring 10 cm on each side. A milliliter is the volume of a cube measuring 1 cm on each side. A capital L is used to signify liter, so it is not confused with the number 1.

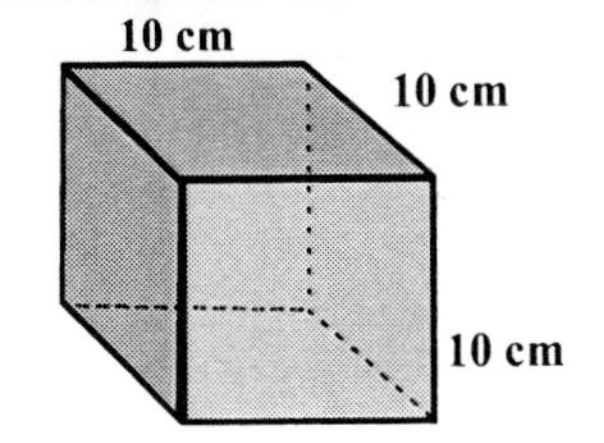

Volume = 1000 cm^3 = 1 liter
(a little more than a quart)

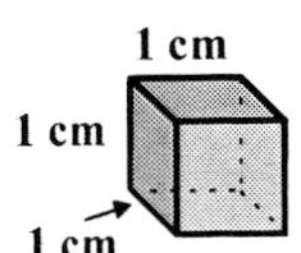

Volume = 1 cm^3 = 1 mL
(an eyedropper holds 1 mL)

6.8 Understanding Grams

The basic unit of **mass** in the metric system is the **gram**. Gram is abbreviated "g".

A large paper clip has a mass of about 1 gram (1 g).

1,000 grams = 1 kilogram (kg) = a little over 2 pounds

1 milligram (mg) = 0.001 gram. This is an extremely small amount and is used in medicine. An aspirin tablet is 300 mg.

6.9 Estimating Metric Measurements

Choose the best estimates.

1. The height of an average man
   (A) 18 cm
   (B) 1.8 m
   (C) 6 km
   (D) 36 mm

2. The volume of a coffee cup
   (A) 300 mL
   (B) 20 L
   (C) 5 L
   (D) 1 kL

3. The width of this book
   (A) 215 mm
   (B) 75 cm
   (C) 2 m
   (D) 1.5 km

4. The weight of an average man
   (A) 5 mg
   (B) 15 cg
   (C) 25 g
   (D) 90 kg

5. The length of a basketball player's foot
   (A) 2 m
   (B) 1 km
   (C) 30 cm
   (D) 100 mm

6. The mass of a dime
   (A) 3 g
   (B) 30 g
   (C) 10 cg
   (D) 1 kg

7. The width of your hand
   (A) 2 km
   (B) 0.5 m
   (C) 25 cm
   (D) 90 mm

8. The length of a basketball court
   (A) 1000 mm
   (B) 250 cm
   (C) 28 m
   (D) 2 km

Choose the best unit of measure.

9. The distance from Atlanta to Macon
   (A) millimeter
   (B) centimeter
   (C) meter
   (D) kilometer

10. The length of a house key
    (A) millimeter
    (B) centimeter
    (C) meter
    (D) kilometer

11. The thickness of a nickel
    (A) millimeter
    (B) centimeter
    (C) meter
    (D) kilometer

12. The width of a classroom
    (A) millimeter
    (B) centimeter
    (C) meter
    (D) kilometer

13. The length of a piece of chalk
    (A) millimeter
    (B) centimeter
    (C) meter
    (D) kilometer

14. The height of a pine tree
    (A) millimeter
    (B) centimeter
    (C) meter
    (D) kilometer

6.10 Converting Units within the Metric System

Converting units such as kilograms to grams or centimeters to decimeters is easy now that you know how to multiply and divide by multiples of ten.

Prefix Meaning	kilo (k) 1000	hecto (h) 100	deka (da) 10	unit (m, L, g) 1	deci (d) 0.1	centi (c) 0.01	milli (m) 0.001

Example 2: 2 L = ____mL

2.000 L = 2000 mL

Look at the chart above. To move from liters to milliliters, you move to the right three places. So, to convert the 2 L to mL, move the decimal point three places to the right. You will need to add three zeros.

Example 3: 5.25 cm = ____m

005.25 cm = 0.0525 m

To move from centimeters to meters, you need to move two spaces to the left. So, to convert 5.25 cm to m, move the decimal point two spaces to the left. Again, you need to add zeros.

Make the following conversions.

1. 35 mg = ______ g
2. 6 km = ______ m
3. 21.5 mL = ______ L
4. 4.9 mm = ______ cm
5. 5.35 kL = ______ mL
6. 32.1 mg = ______ kg
7. 156.4 m = ______ km
8. 25 mg = ______ cg
9. 17.5 L = ______ mL
10. 4.2 g = ______ kg
11. 0.06 daL = ______ dL
12. 0.417 kg = ______ cg
13. 18.2 cL = ______ L
14. 81.2 dm = ______ cm
15. 72.3 cm = ______ m
16. 0.003 kL = ______ L
17. 5.06 g = ______ mg
18. 1.058 mL = ______ cL
19. 43 hm = ______ km
20. 2.057 m = ______ cm
21. 564.3 g = ______ kg

Chapter 6 Review

Fill in the blanks below with the appropriate unit of measurement.

1. A box of assorted chocolates might weigh about 1 _______ (English).
2. A compact disc is about 7 _______ (English) across.
3. Soft drinks are sold in _______ (metric).
4. A vitamin C tablet has a mass of 500 _______(metric).

Fill in the blanks below with the appropriate English or metric conversions.

5. Two gallons equals _______ cups.
6. 4.2 L equals _______ mL.
7. $3\frac{1}{2}$ yards equals _______ inches.
8. 6,800 m equals _______ kilometers.
9. 36 oz. equals _______ pounds.
10. 730 mg equals _______ kg.

Solve the following problems.

11. 120 m = ______ km
12. 9 g = ______ mg
13. 0.02 kL = ______ L
14. 1.5 mg = ______ g
15. 15 cm = ______ mm
16. 5 L = ______ mL
17. 0.005 kg = ______ g
18. 55 mL = ______ L
19. 30 cm = ______ m

Chapter 6 Test

1. The length of a standard pencil is approximately ____ inches.

   A. 7
   B. 2
   C. 14
   D. 80

2. The height of a standard door is approximately ____ feet.

   A. 17
   B. 18
   C. 2
   D. 7

3. Convert: 42 inches = ____ feet

   A. 3
   B. 3.5
   C. 504
   D. 400

4. Convert: 5 miles = ____ yards

   A. 60
   B. 5,000
   C. 8,800
   D. 15

5. A ruler is approximately ____ centimeters in length.

   A. 6
   B. 15
   C. 28
   D. 36

6. The diameter of a grapefruit is approximately ____ cm.

   A. 10
   B. 15
   C. 2
   D. 20

7. Convert: 62 km = ____ m

   A. 620
   B. 62,000
   C. 6,200
   D. 0.062

8. Convert: 56 mm = ____ cm

   A. 560
   B. 5,600
   C. 5.6
   D. 0.56

Chapter 7
Plane Geometry

This chapter covers the following Georgia Performance Standards:

M5M	Measurement	M5M1.a, b, c, d, e, f
M5G	Geometry	M5G1 M5G2
M5P	Process Skills	M5P1.c

7.1 Types of Triangles

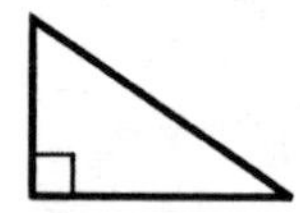

right triangle
contains 1 right ∠

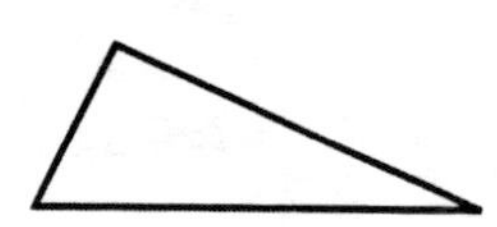

acute triangle
all angles are acute
(less than 90°)

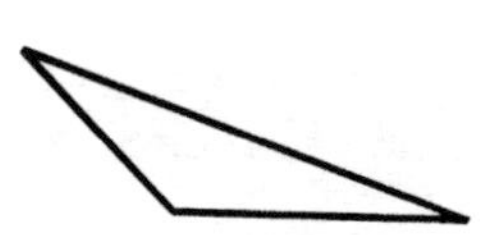

obtuse triangle
one angle is obtuse
(greater than 90°)

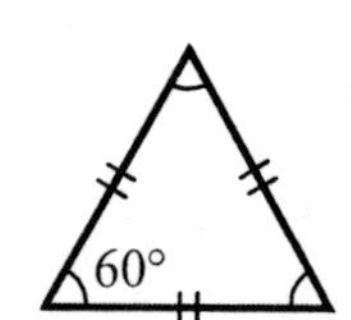

equilateral triangle
all three sides equal
all angles are 60°

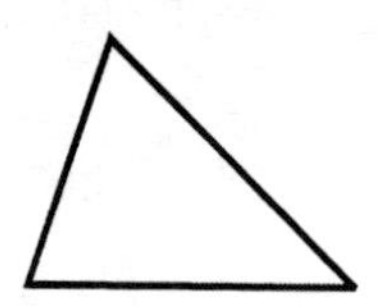

scalene triangle
no sides equal
no angles equal

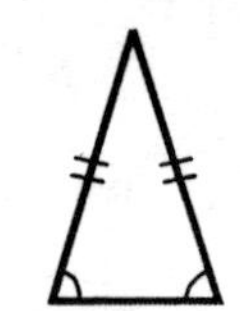

isosceles triangle
two sides equal
two angles equal

7.2 Types of Polygons

square
equal sides
90° ∠s

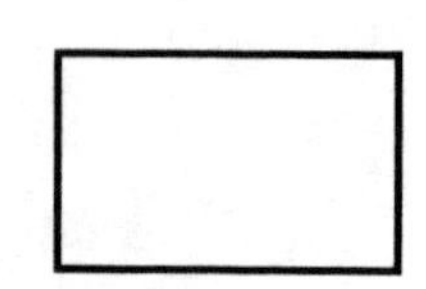

rectangle
opposite sides
parallel, 90° ∠s

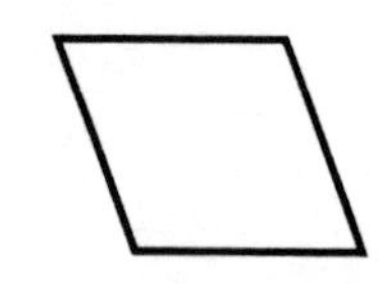

parallelogram
opposite sides
parallel

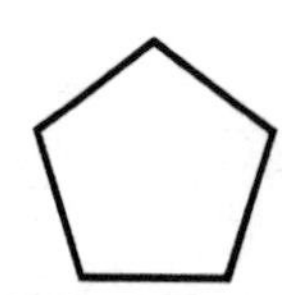

pentagon
5 sides

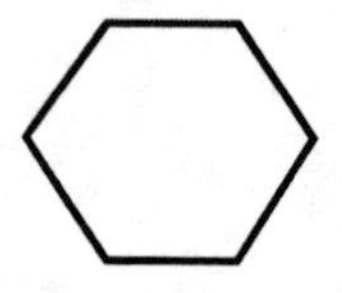

hexagon
6 sides

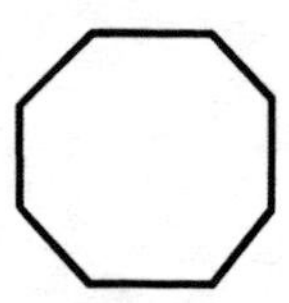

octagon
8 sides

7.3 Estimating Area

Example 1: Estimate the area of the figure below.

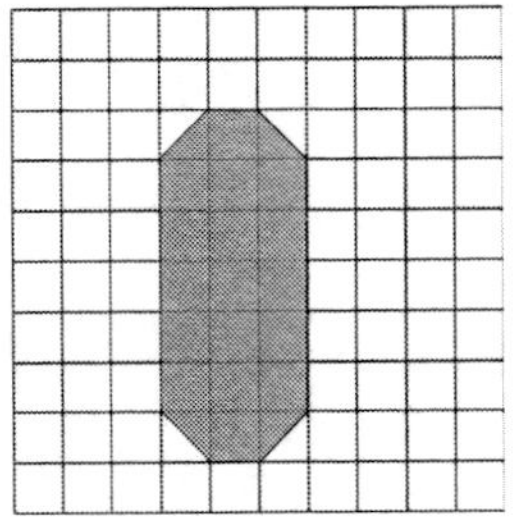

Step 1: Count the number of whole boxes inside the figure. There are 17 whole boxes inside the figure.

Step 2: Now count the number of half boxes inside the figure. There are 4 half boxes inside the figure. Because it takes 2 halves to make a whole, we will divide our 4 half boxes by 2.
$4 \div 2 = 2$

Step 3: Now that we know our 4 half boxes are equal to 2 whole boxes, we can add these 2 whole boxes to our original 17.
$2 + 17 = 19$
This figure has an area of 19 units^2.

Estimate the area of the following figures.

1. 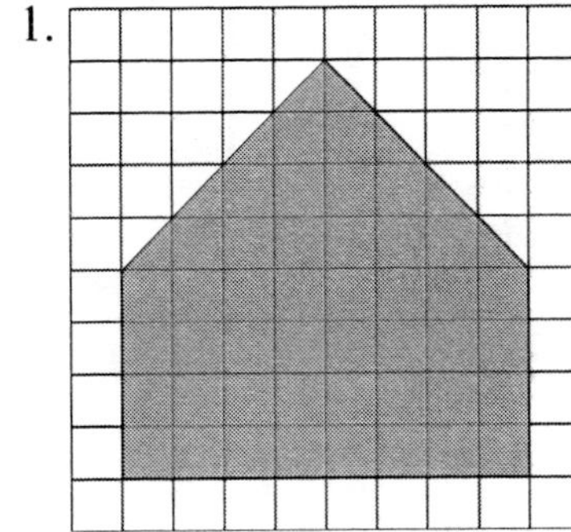

2. 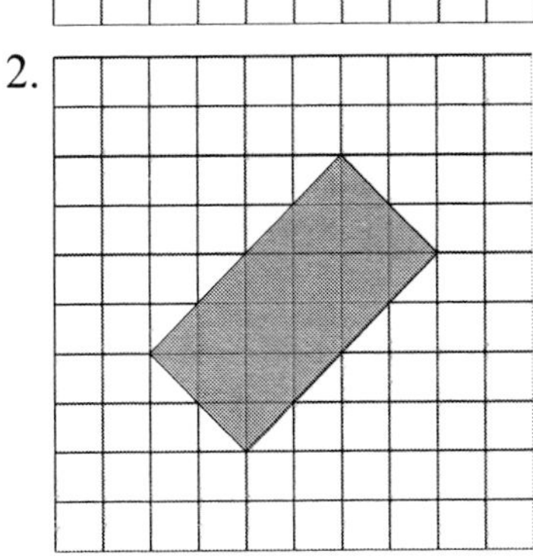

3. 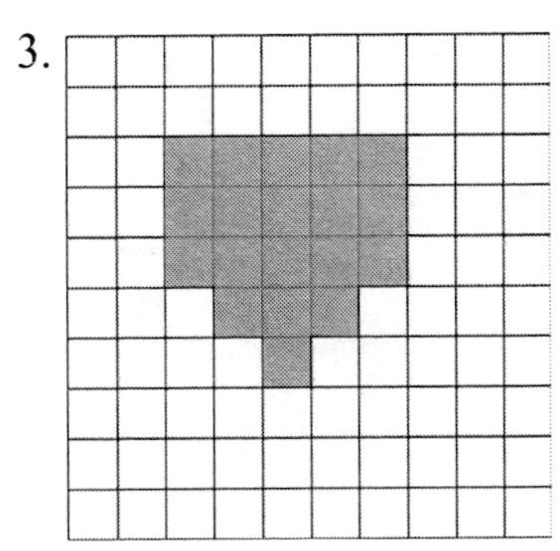

4. 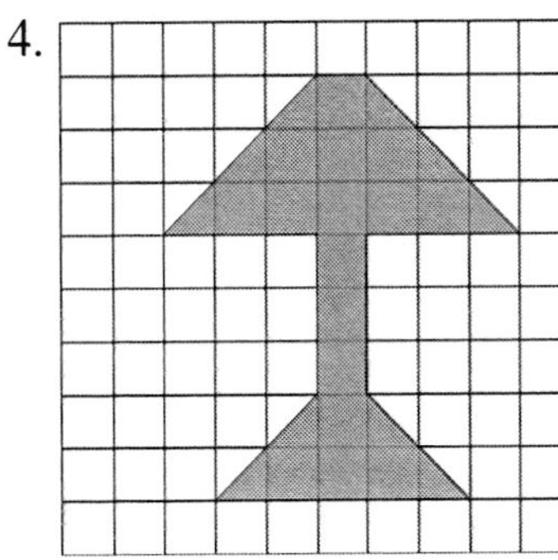

7.4 Area of Squares and Rectangles

Area is always expressed in square units, such as in^2, m^2, and ft^2.

The area, (A), of squares and rectangles equals length (l) times width (w). $A = l \times w$.

Example 2:

4 cm

4 cm

$A = lw$
$A = 4 \times 4$
$A = 16$ $\textbf{cm}^2$

If a square has an area of 16 cm², it means that it will take 16 squares that are 1 cm on each side to cover the area that is 4 cm on each side.

Find the area of the following squares and rectangles using the formula $A = lw$.

1. 10 ft, 10 ft

2.

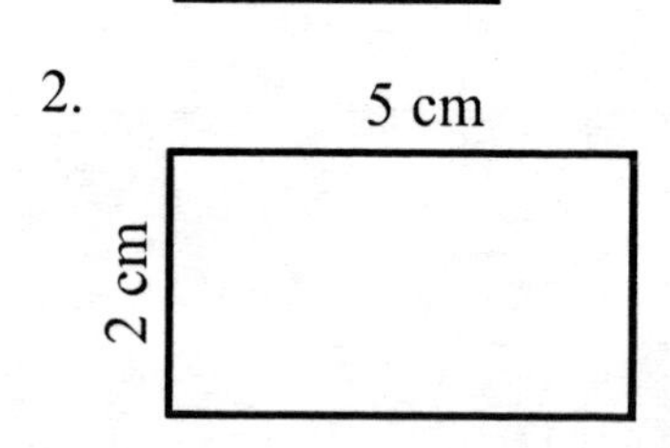

3.

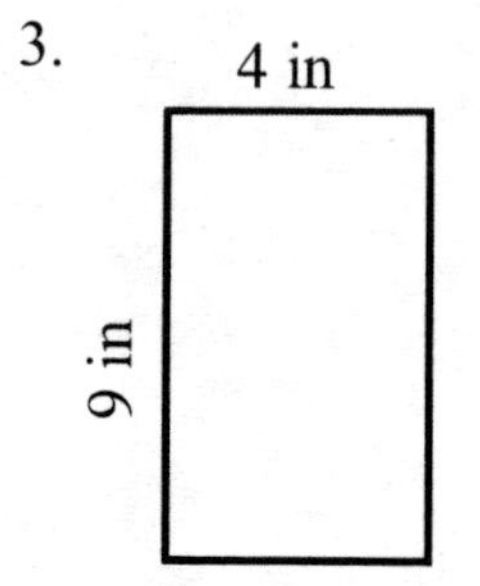

4. 9 in, 20 in

5. 6 ft, 6 ft

6.

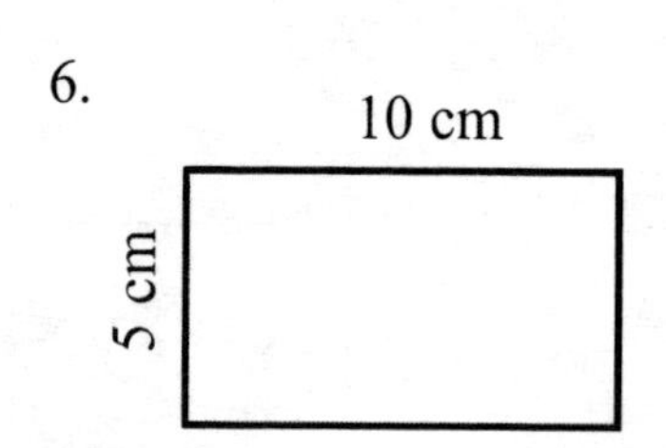

7. 4 ft, 2 ft

8. 5 in, 8 in

9.

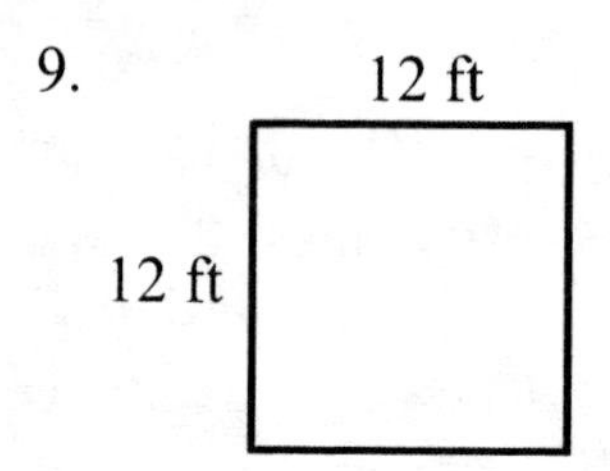

10.

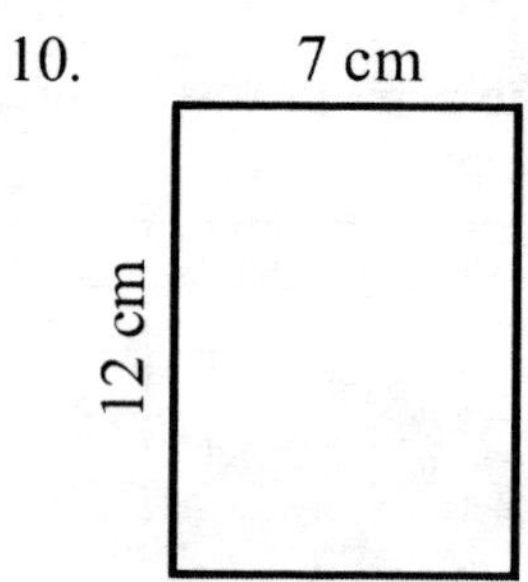

11.

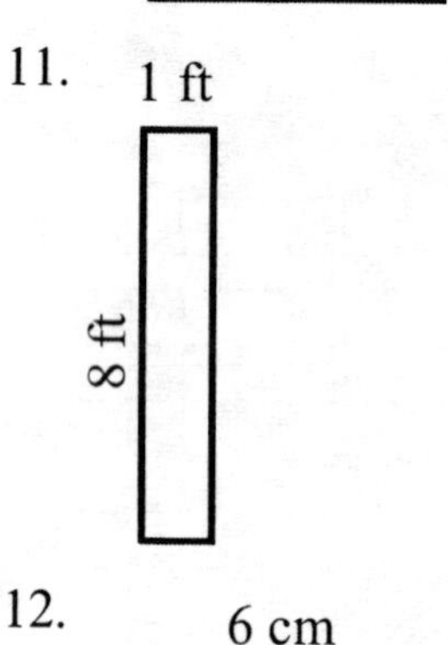

12. 6 cm, 7 cm

7.5 Deriving the Area of a Parallelogram

A parallelogram is a 4-sided shaped with opposite sides parallel. If you were to cut off a triangle from the side of a parallelogram, it would fit on the other side of the parallelogram and make a rectangle. So you can see how finding the area of a parallelogram is the same as finding the area of a rectangle, base × height.

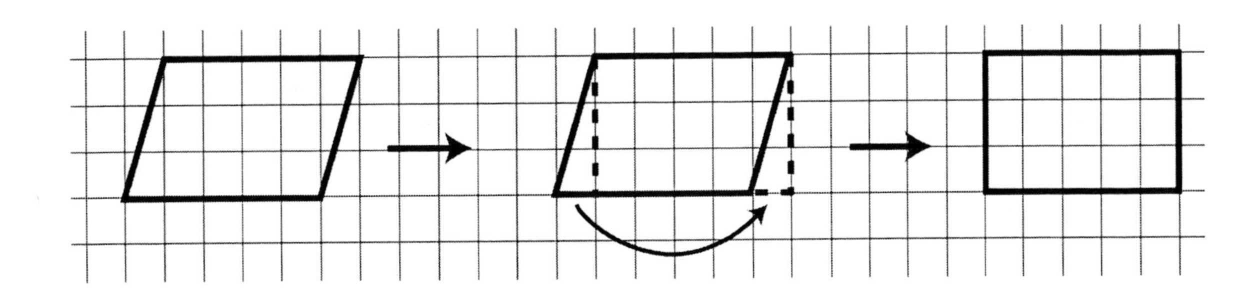

Example 3: Find the area of the parallelogram below.

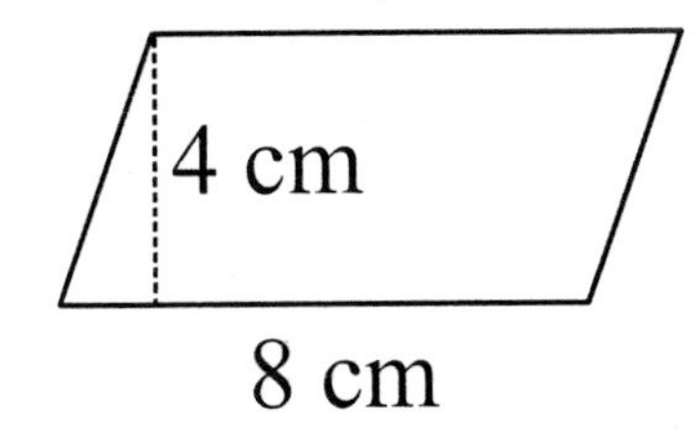

Step 1: The formula for finding the area of a parallelogram is $A = bh$, where b = base and h = height.

Plug the known variables into the equation.

We know that $h = 4$ cm and $b = 8$ cm.

$A = bh$

$A = 8 \times 4$

Step 2: Solve the equation.

$A = 8 \times 4 = 32 \text{ cm}^2$

The area of the parallelogram is 32 cm^2.

7.6 Area of Parallelograms

Example 4: Find the area of the following parallelogram.

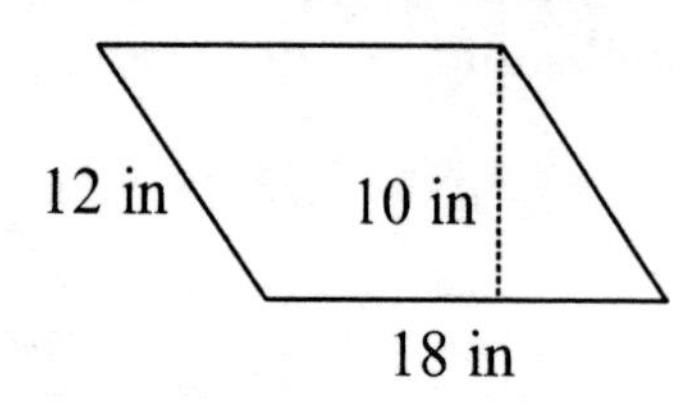

The formula for the area of a parallelogram is $A = bh$.
A = area
b = base
h = height

Step 1: Insert measurements from the parallelogram into the formula: $A = 18 \times 10$.

Step 2: Multiply. $18 \times 10 = 180 \text{ in}^2$

Find the area of the following parallelograms and trapezoids.

1. 12 in

13 in

11 in

1452

3. 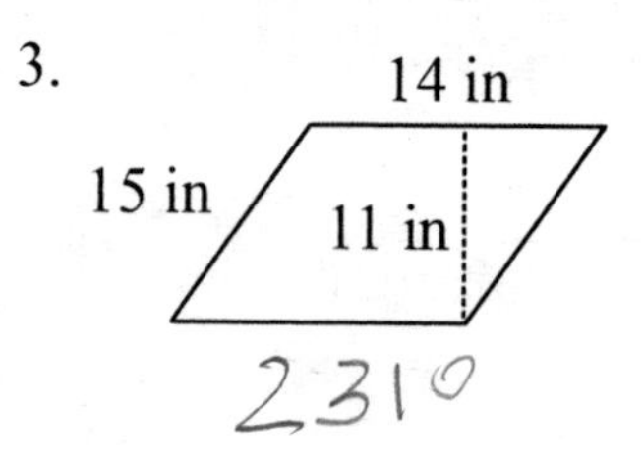

2310

2. 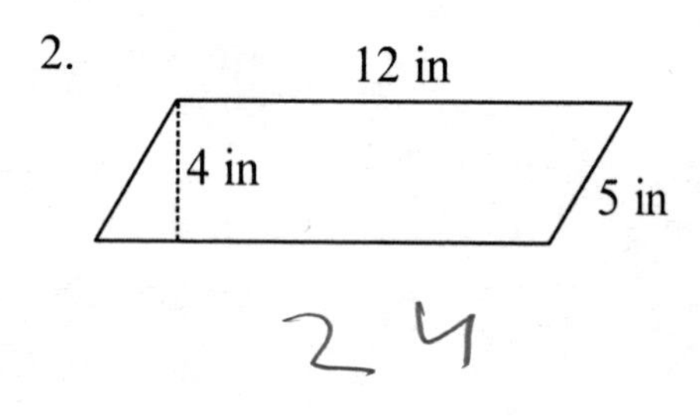

24

4. 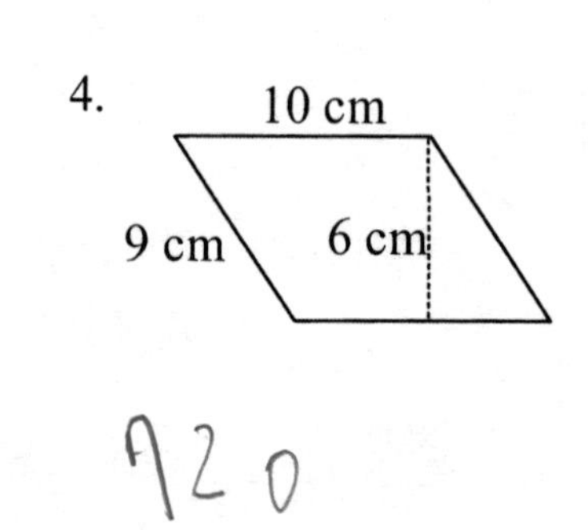

920

7.7 Deriving the Area of a Triangle

If you cut a rectangle in half diagonally, you get two congruent (same size and shape) triangles. If the area of a rectangle is base $\times$ height, then the area of those triangles you cut would be half as much as the rectangle. The area of a triangle is $\frac{1}{2}\times$ base $\times$ height. This formula works for any shape triangle.

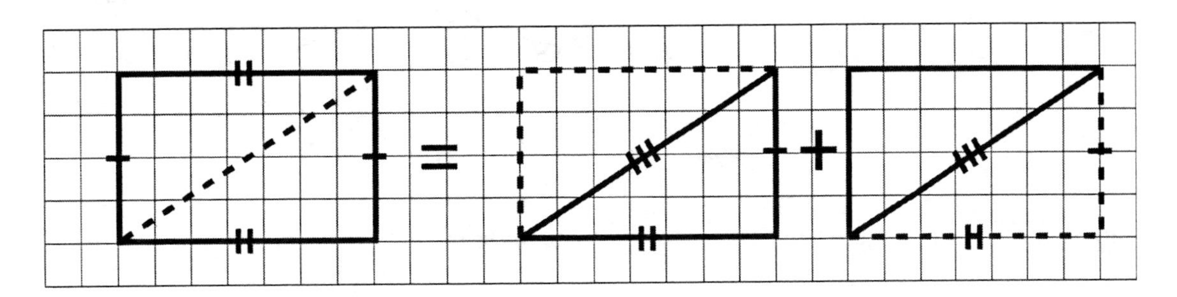

Example 5: Find the area of the triangle below.

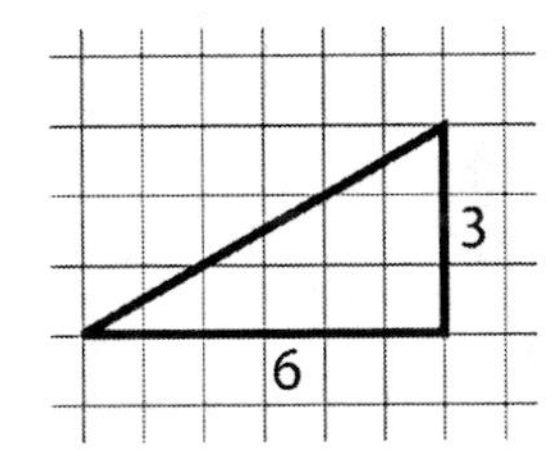

Step 1: Plug values into equation.
$A = \frac{1}{2}bh$, where $b = 6$ and $h = 3$
$A = \frac{1}{2} \times 6 \times 3$

Step 2: Solve.
$A = \frac{1}{2} \times 6 \times 3 = 3 \times 3 = 9$

Example 6: Find the area of the triangle below.

9 m
6 m
height
14 m

Step 1: Plug values into equation.
$A = \frac{1}{2}bh$, where $b = 14$ and $h = 6$
$A = \frac{1}{2} \times 14 \times 6$

Step 2: Solve.
$A = \frac{1}{2} \times 14 \times 6 = 7 \times 6 = 42$

7.8 Area of Triangles

Example 6: Find the area of the following triangle.

The formula for the area of a triangle is as follows:

$$A = \frac{1}{2} \times b \times h$$

$A =$ area
$b =$ base
$h =$ height or altitude

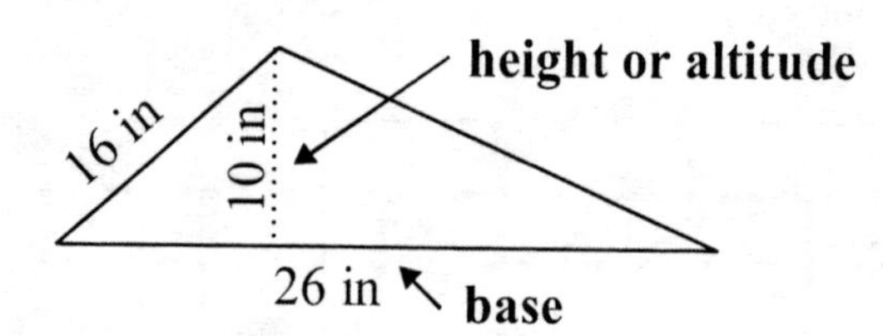

Step 1: Insert the measurements from the triangle into the formula: $A = \frac{1}{2} \times 26 \times 10$

Step 2: Cancel and multiply. $A = \frac{1}{\cancel{2}_{1}} \times \frac{\cancel{26}^{13}}{1} \times \frac{10}{1} = 130 \text{ in}^2$

Note: **Area is always expressed in square units such as in^2, ft^2, or m^2.**

Find the area of the following triangles. Remember to include units.

1. 3 in, 5 in, 4 in

2. 7 cm, 6 cm ← height, 12 cm

3. 6 ft, 9 ft, 7 ft

4. 12 cm, 12 cm

5. 3 ft, 2 ft

6. 20 cm, 16 cm

7. 8 m, height → 7 m, 15 m

8.

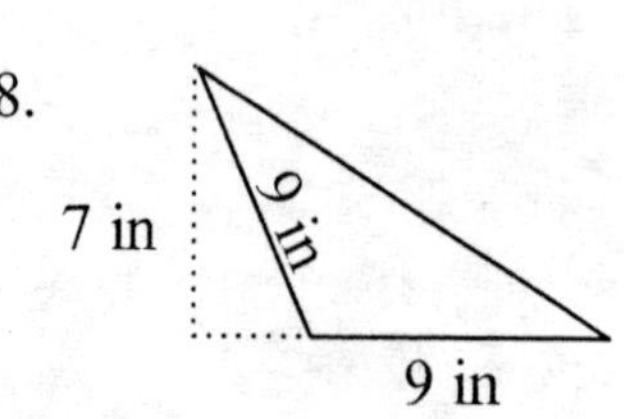

9. 2 ft, 2 ft

10. 5 ft, 4 ft, 6 ft

11.

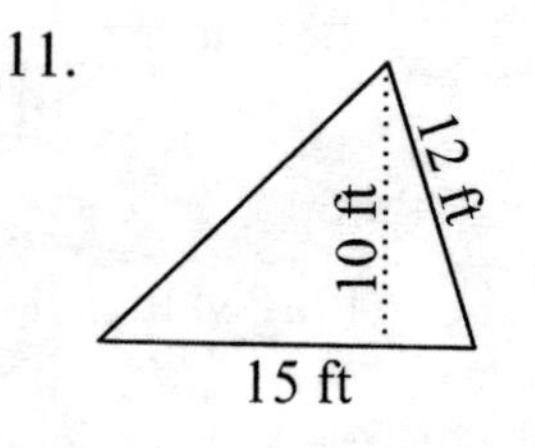

12.

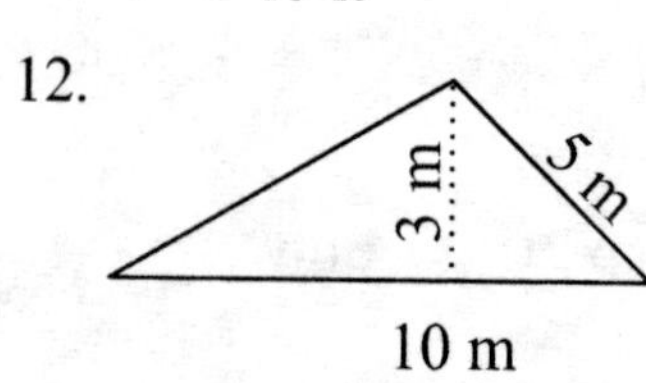

7.9 Circumference

Circumference, C - the distance around the outside of a circle
Diameter, d - a line segment passing through the center of a circle from one side to the other
Radius, r - a line segment from the center of a circle to the edge of a circle
Pi, π- the ratio of a circumference of a circle to its diameter. $\pi = 3.14$

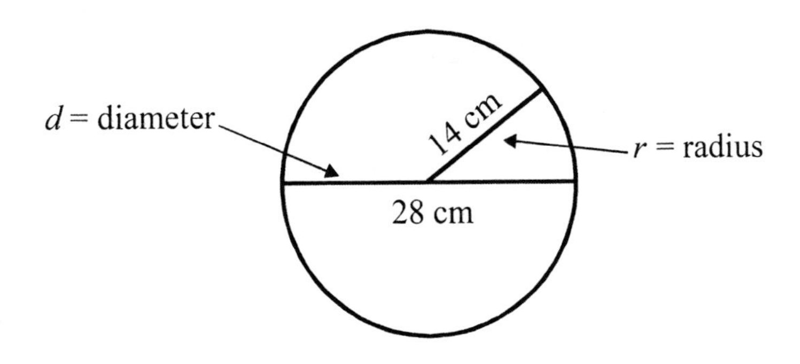

The formula for the circumference of a circle is $C = 2\pi r$ or $C = \pi d$. (The formulas are equal because the diameter is equal to twice the radius, $d = 2r$.)

Example 7: Find the circumference of the circle above.

$C = \pi d$ Use $\pi = 3.14$	$C = 2\pi r$
$C = 3.14 \times 28$	$C = 2 \times 3.14 \times 14$
$C = 87.92$ cm	$C = 87.92$ cm

Use the formulas given above to find the circumferences of the following circles. Use $\pi = 3.14$.

1. 8 in — $C =$ ______

2. 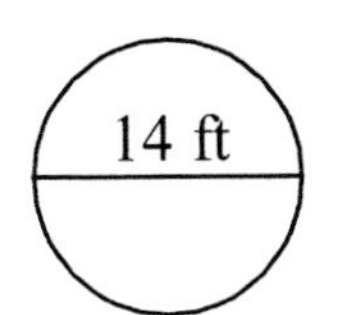

$C =$ ______

3. 2 cm — $C =$ ______

4. 6 m — $C =$ ______

5. 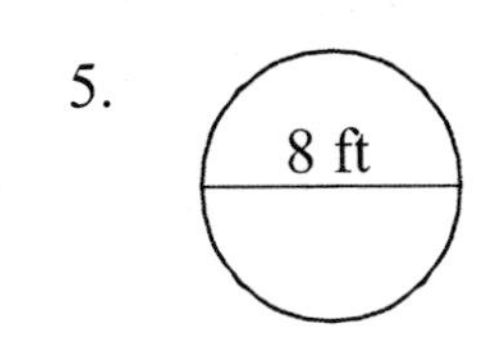

$C =$ ______

6. 3 ft — $C =$ ______

7. 12 in — $C =$ ______

8. 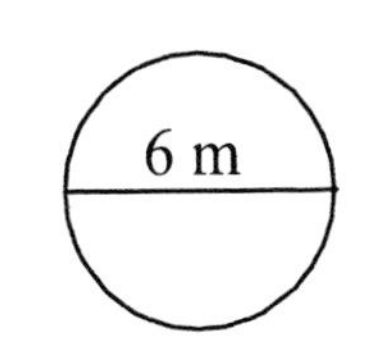

$C =$ ______

9. 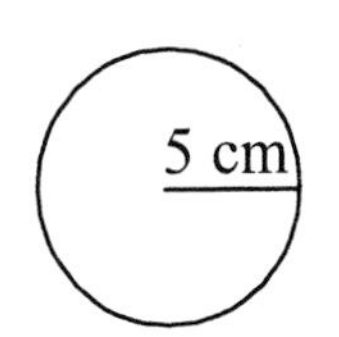

$C =$ ______

10. 16 in — $C =$ ______

7.10 Comparing Diameter, Radius, and Circumference in a Circle

Circumference, C, is the distance around the outside of a circle. ($C = 2r$ or $C = d$)

Diameter, d, is a line segment passing through the center of a circle from one side to the other.

Radius, r, is a line segment from the center of a circle to the edge of the circle.

Pi, π, is the ratio of the circumference of a circle to its diameter. $\left(\pi = 3.14 \text{ or } \frac{22}{7}\right)$

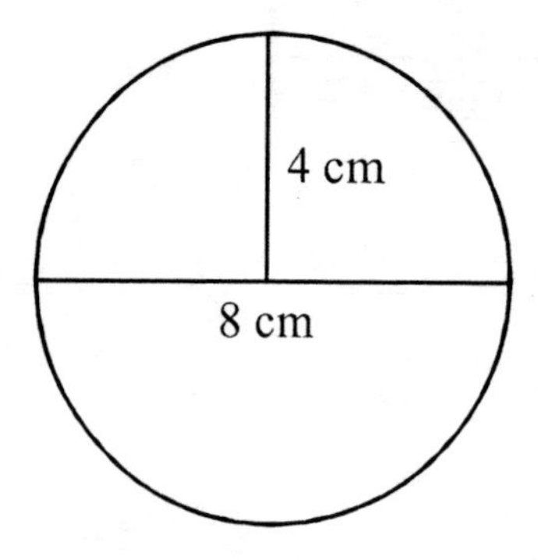

Answer the following questions about the circle above.

1. What is radius of the circle?

2. What is the diameter of the circle?

3. Pi is the ratio of which two parts of a circle?

4. What is the circumference of the circle?

5. What is the symbol for Pi?

6. How many different ways can you draw the radius of the circle?

7. Does the diameter have to pass through the center of the circle?

8. Pi can be written as 3.14. How else can it be written?

7.11 Area of a Circle

The area is how many square units of measure would fit inside a circle. The formula for the area of a circle is $A = \pi r^2$. r^2 means $r \times r$.

Example 8: Find the area of the circle, using $\pi = 3.14$.

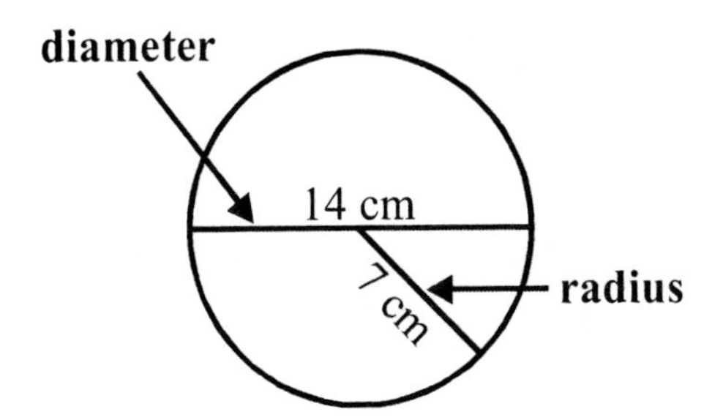

Let $\pi = 3.14$
$A = \pi r^2 = \pi r \times r$
$A = 3.14 \times 7^2 = 3.14 \times 7 \times 7$
$A = 3.14 \times 49$
$A = 153.86 \text{ cm}^2$

Find the area of the following circles. Remember to include units.

$\pi = 3.14$

1. 5 in — $A =$ ______
2. 16 ft — $A =$ ______
3. 8 cm — $A =$ ______
4. 3 m — $A =$ ______

Fill in the chart below. Include appropriate units.

			Area
	Radius	Diameter	$\pi = 3.14$
5.	9 ft	18	
6.		4 in	
7.	8 cm		
8.		20 ft	
9.	14 m		
10.		18 cm	
11.	12 ft		
12.		6 in	

7.12 Two-Step Area Problems

Solving the following problems will require two steps. You will need to find the area of two figures, and then either add or subtract the two areas to find the answer.

Example 9:
Find the area of the living room below.

Figure 1

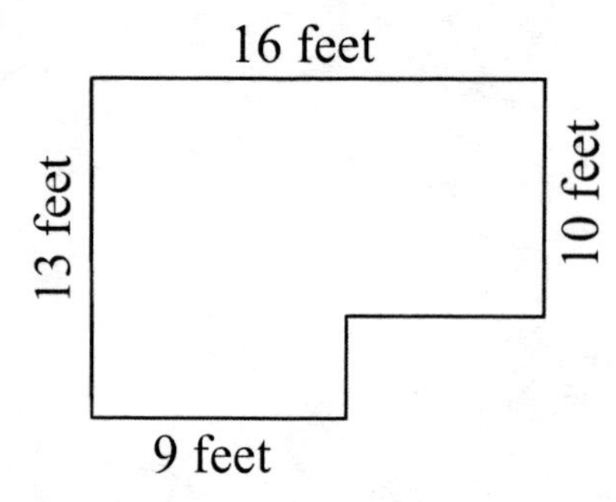

Step 1: Complete the rectangle as in Figure 2, and compute the area as if it were a complete rectangle.

Figure 2

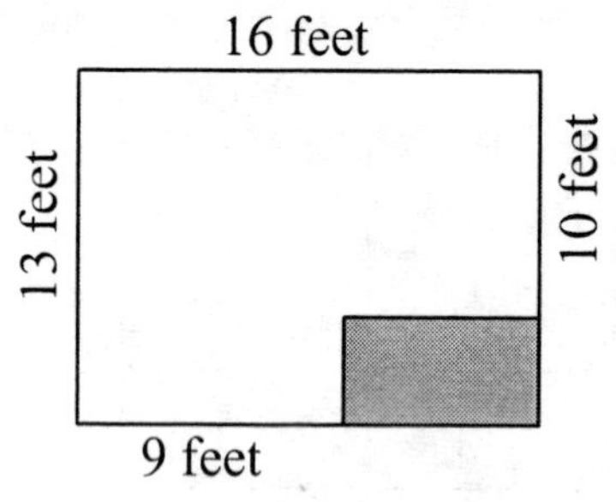

$$A = \text{length} \times \text{width}$$
$$A = 16 \times 13$$
$$A = 208 \text{ ft}^2$$

Step 2: Figure the area of the shaded part.

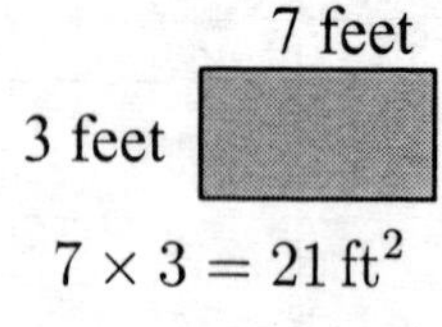

$$7 \times 3 = 21 \text{ ft}^2$$

Step 3: Subtract the area of the shaded part from the area of the complete rectangle.

$$208 - 21 = 187 \text{ ft}^2$$

Example 10:
Find the area of the shaded sidewalk.

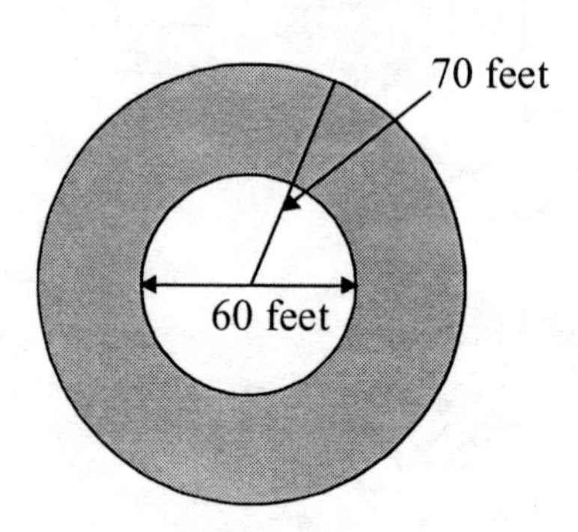

Step 1: Find the area of the outside circle.

$$\pi = 3.14$$
$$A = 3.14 \times 70 \times 70$$
$$A = 15,386 \text{ ft}^2$$

Step 2: Find the area of the inside circle.

$$\pi = 3.14$$
$$A = 3.14 \times 30 \times 30$$
$$A = 2,826 \text{ ft}^2$$

Step 3: Subtract the area of the inside circle from the area of the outside circle.

$$15,386 - 2,826 = 12,560 \text{ ft}^2$$

Find the area of the following figures.

1.

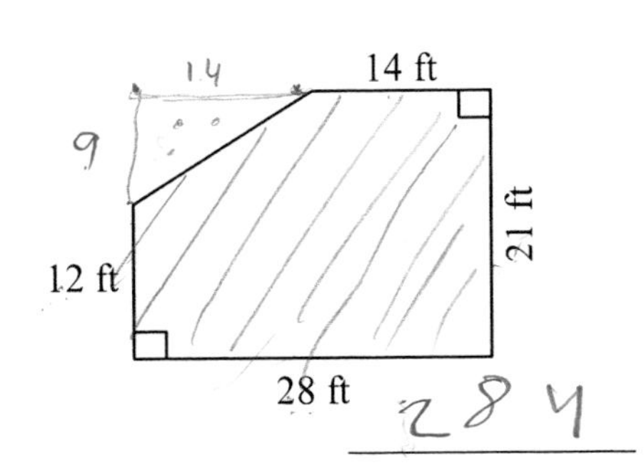

2.

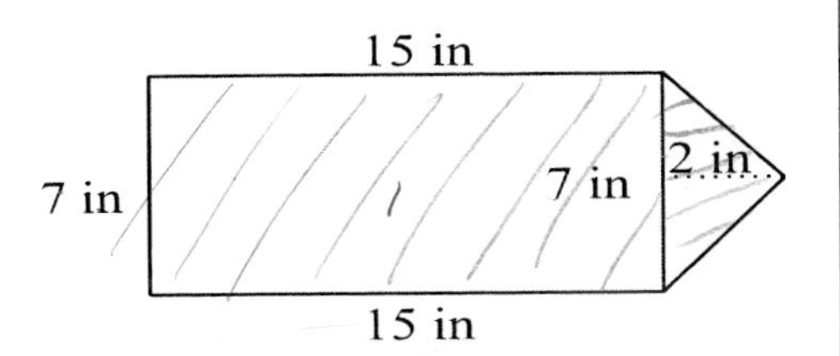

3. What is the area of the shaded circle? Use $\pi = 3.14$, and round the answer to the nearest whole number.

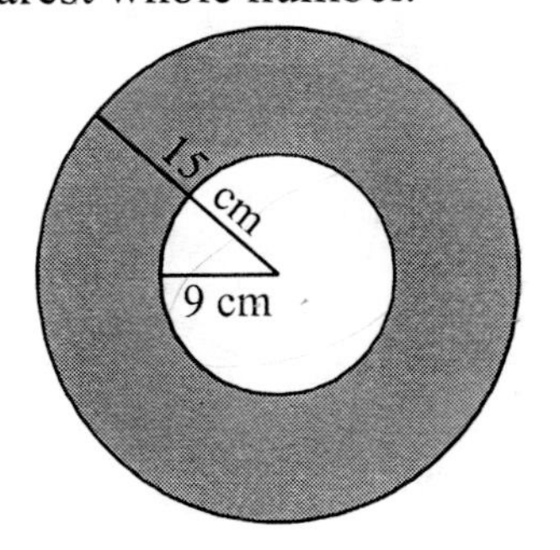

4.

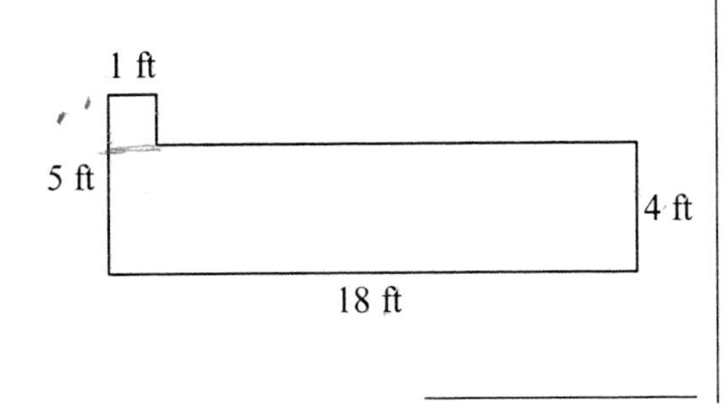

5.

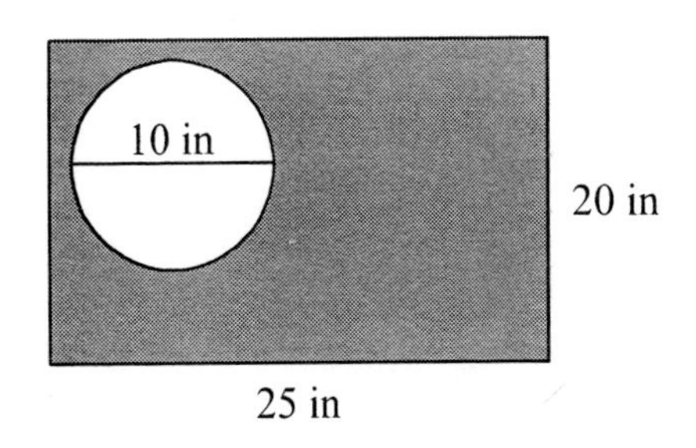

6. What is the area of the shaded part?

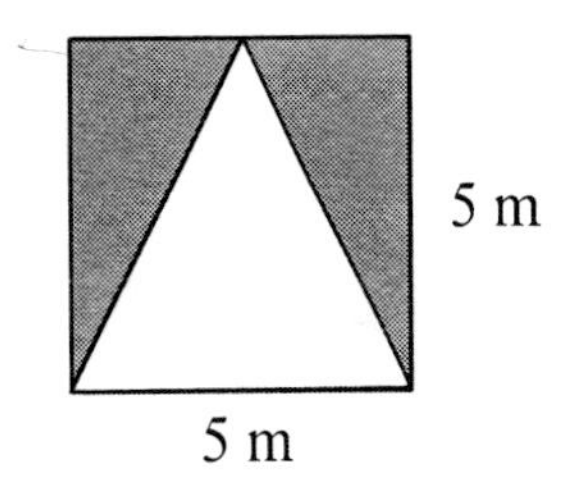

7. What is the area of the shaded part?

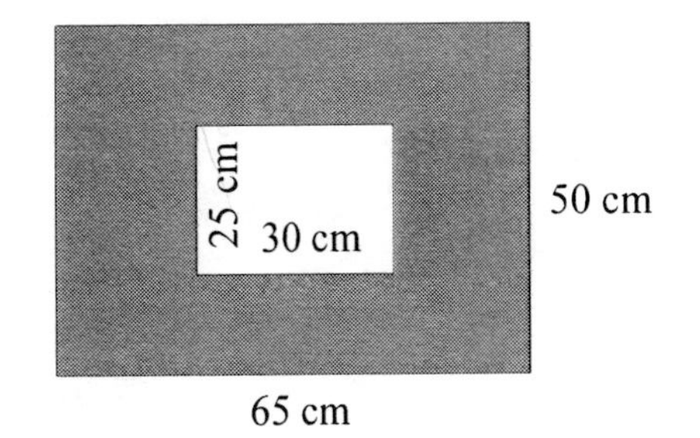

8.

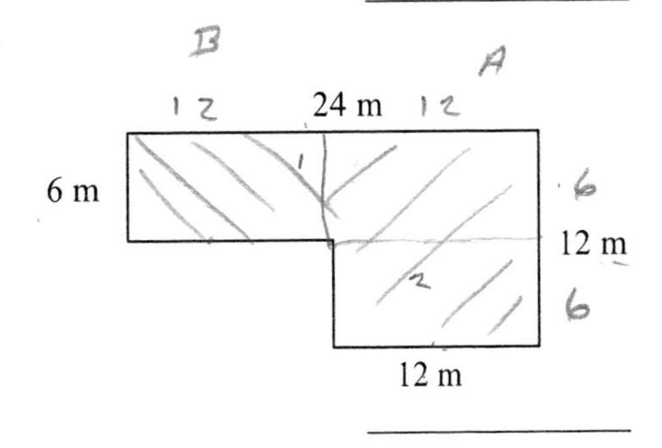

7.13 Similar and Congruent

Similar figures have the same shape but are two different sizes. Their corresponding sides are proportional. **Congruent figures** are exactly alike in size and shape and their corresponding sides are equal. See the examples below.

SIMILAR **CONGRUENT**

Label each pair of figures below as either S if they are similar, C if they are congruent, or N if they are neither.

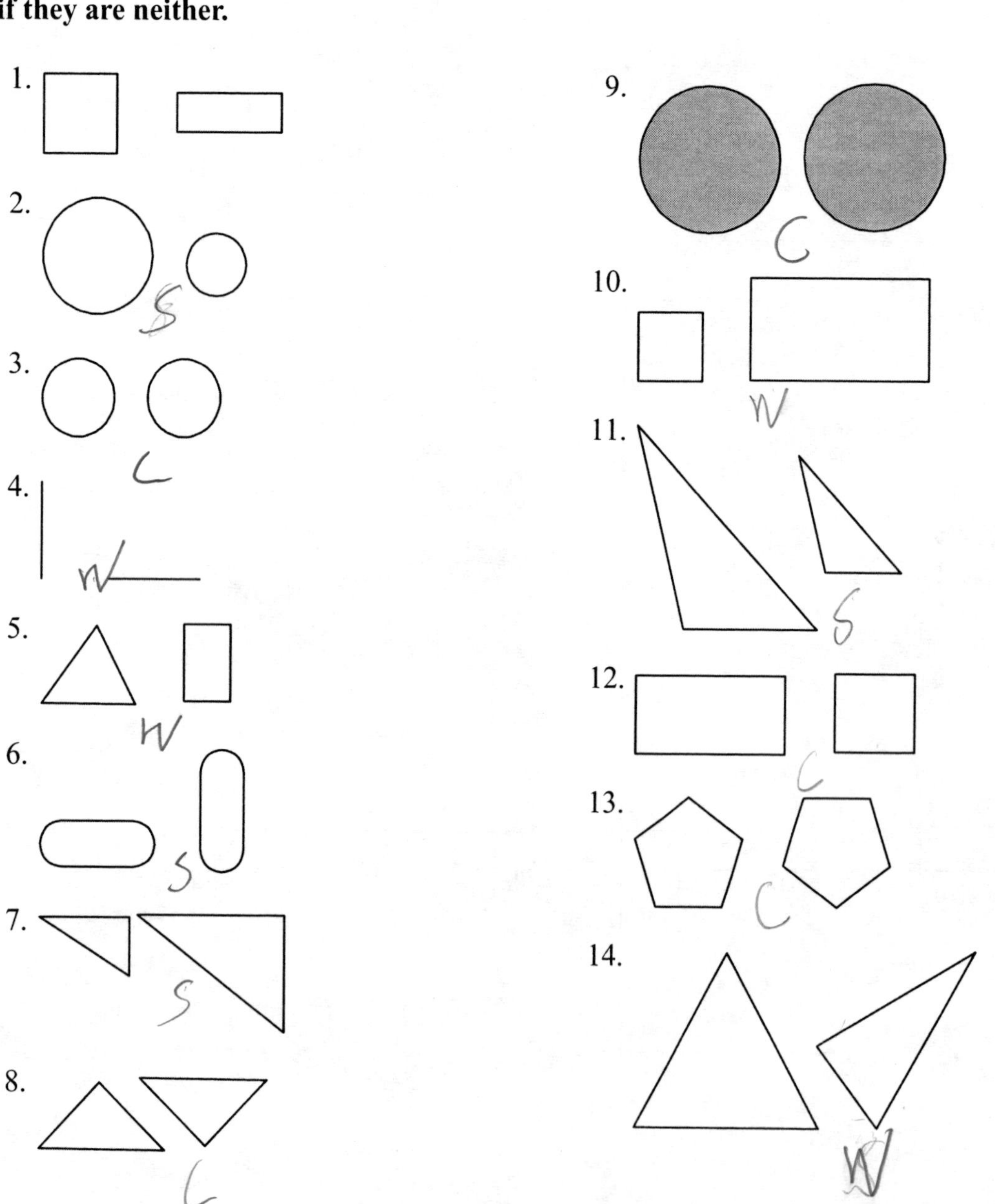

7.14 Congruent Figures

Two figures are **congruent** when they are exactly the same size and shape. If the corresponding sides and angles of two figures are congruent, then the figures themselves are congruent. For example, look at the two triangles below.

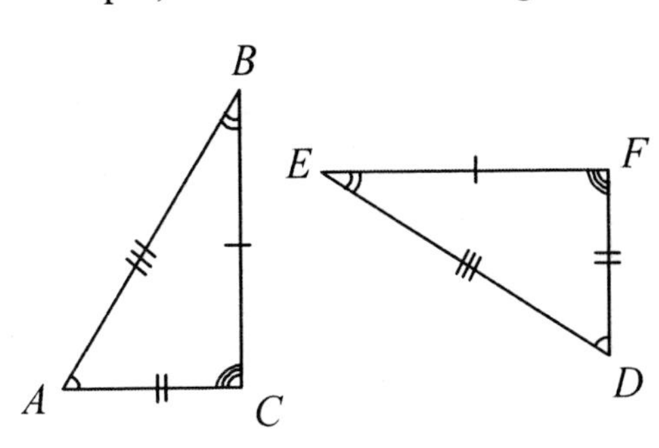

Compare the lengths of the sides of the triangles. The slash marks indicate that $\overline{AB}$ and $\overline{ED}$ have the same length. Therefore, they are congruent, which can be expressed as $\overline{AB} \cong \overline{ED}$. We can also see that $\overline{BC} \cong \overline{EF}$ and $\overline{AC} \cong \overline{FD}$. In other words, the corresponding sides are congruent. Now, compare the corresponding angles. The markings show that the corresponding angles have the same measure and are, therefore, congruent: $\angle A \cong \angle D$, $\angle B \cong \angle E$, and $\angle C \cong \angle F$. Because the corresponding sides and angles of the triangles are congruent, we say that the triangles are congruent: $\triangle ABC \cong \triangle DEF$.

Example 11: Decide whether the figures in each pair below are congruent or not.

PAIR 1

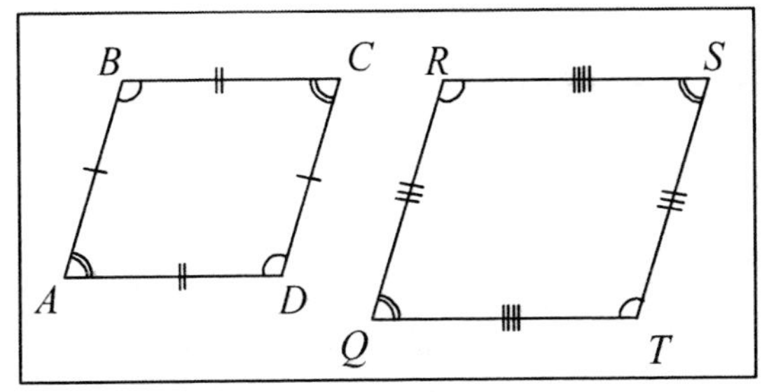

In Pair 1, the two parallelograms have congruent corresponding angles. However, because the corresponding sides of the parallelogram are not the same size, the figures are not congruent.

PAIR 2

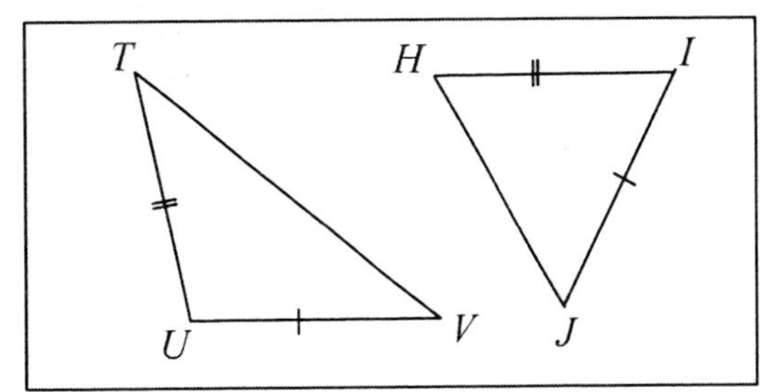

In Pair 2, the two triangles have two corresponding sides which are congruent. However, the hypotenuse of these triangles are not congruent (indicated by the lack of a triple hash mark).

PAIR 3

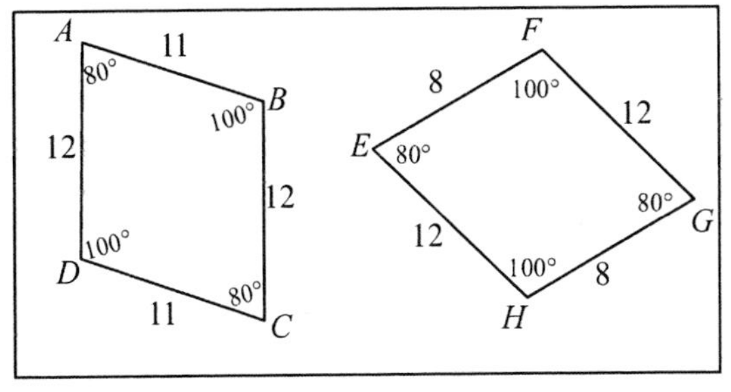

In Pair 3, all of the corresponding angles of these parallelograms are congruent; however, the corresponding sides are not congruent. Therefore, these figures are not congruent.

PAIR 4

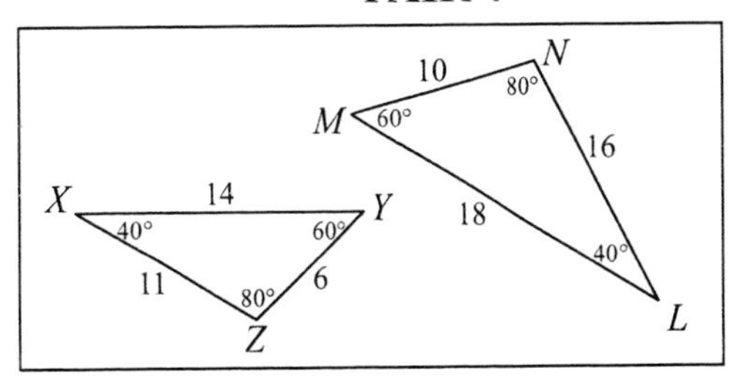

In Pair 4, the triangles share congruent corresponding angles, but the measures for all three corresponding sides of the triangles are not congruent. Therefore, the triangles are not congruent.

Examine the pairs of corresponding figures below. On the first line below the figures, write whether the figures are congruent or not congruent. On the second line, write a brief explanation of how you chose your answer.

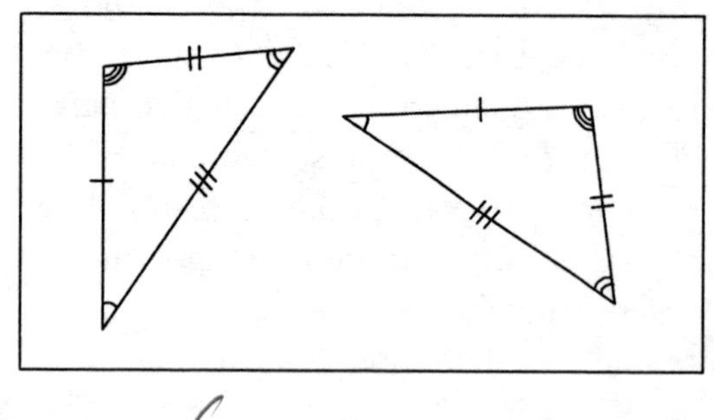

1. c

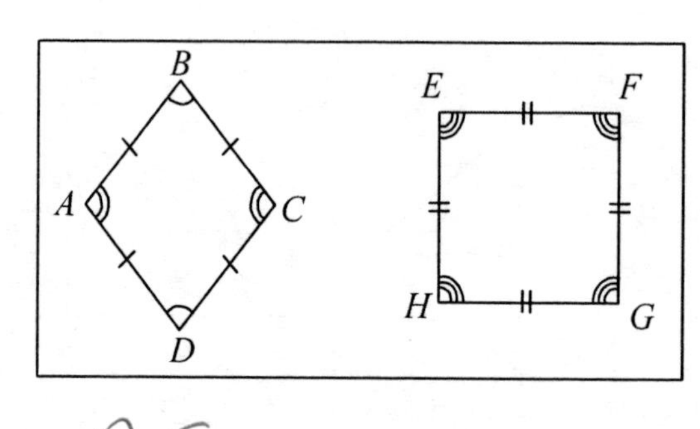

2. nc

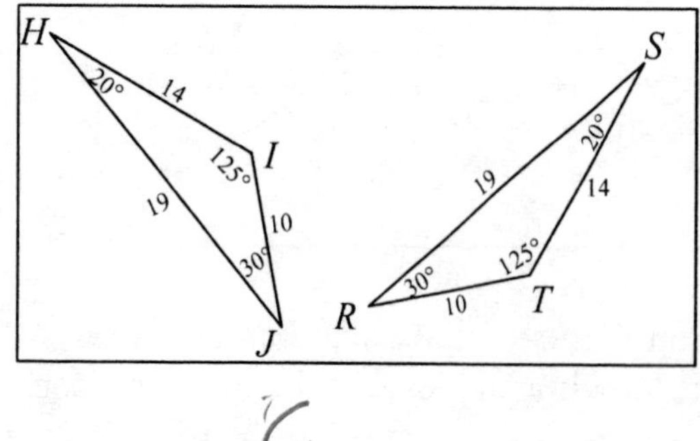

3. c

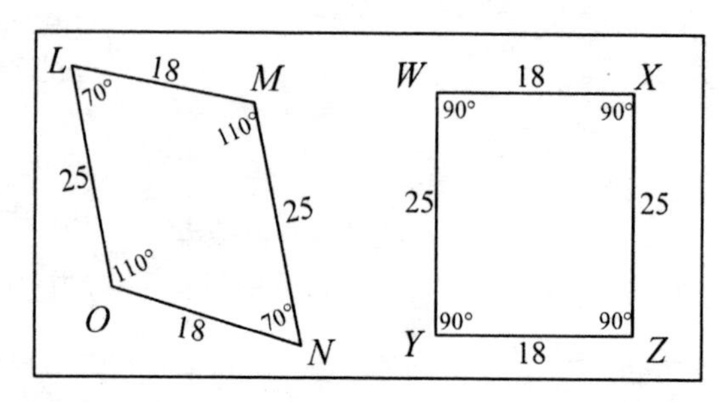

4. nc

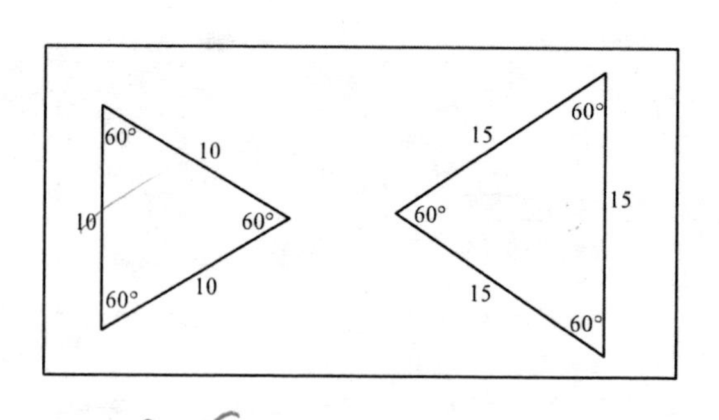

5. nc

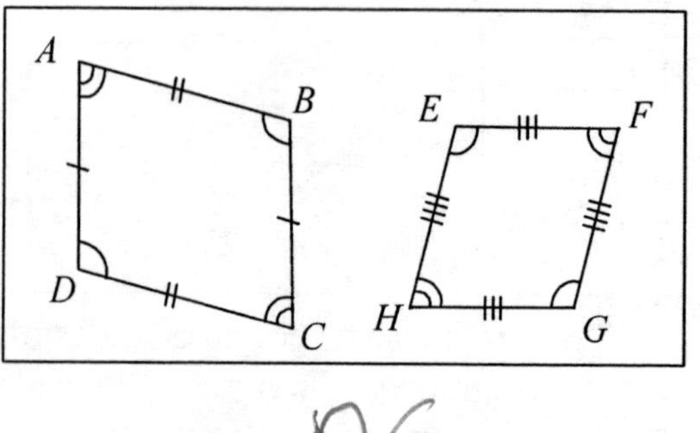

6. nc

Chapter 7 Review

1. Estimate the area of the following figure.

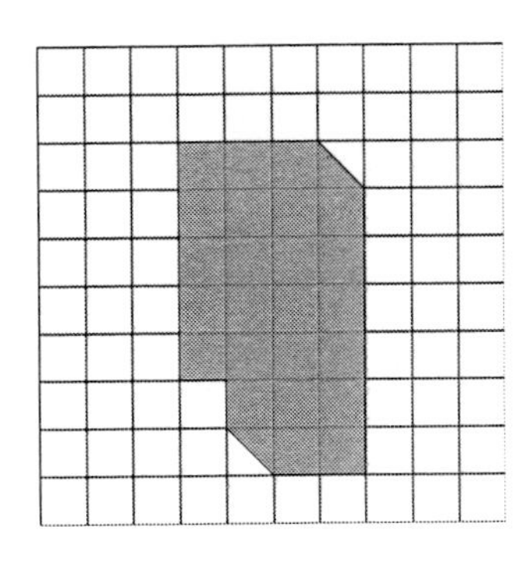

25

2. Estimate the area of the following figure.

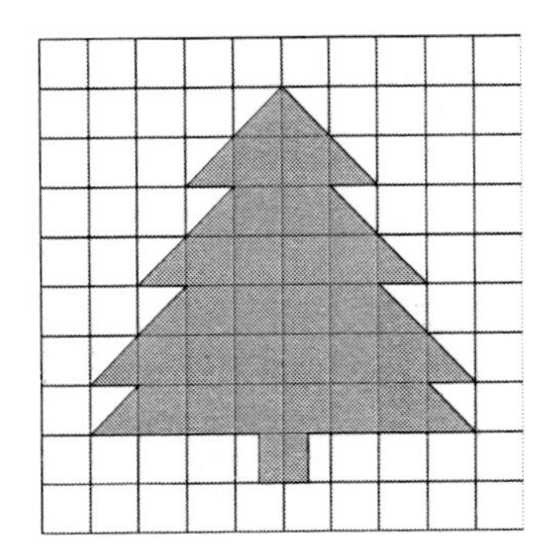

25

3. Calculate the circumference and the area of the following circle. Use $\pi = 3.14$.

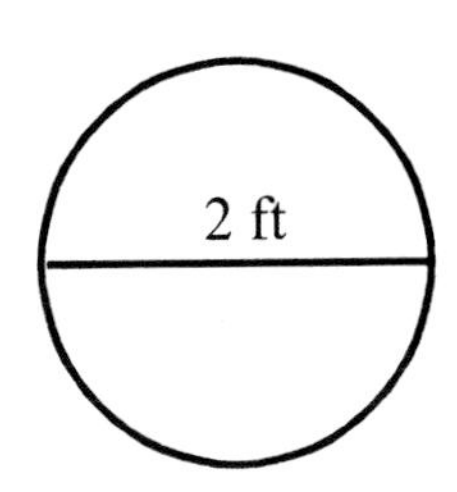

4. Find the area of the shaded region of the figure below.

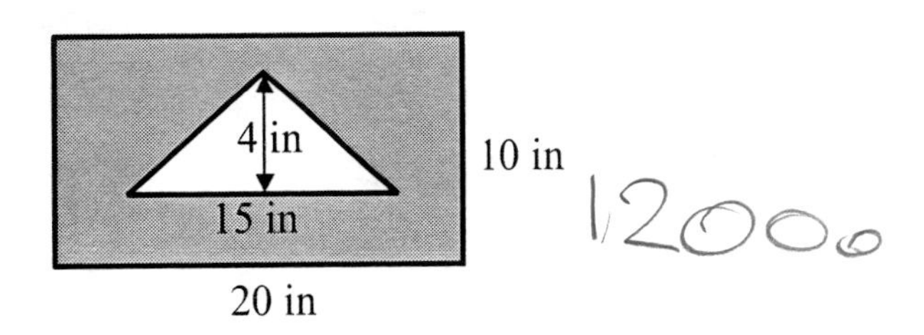

5. Find the area of the shaded part.

4 cm
10 cm

10

6. Calculate the area of the parallelogram.

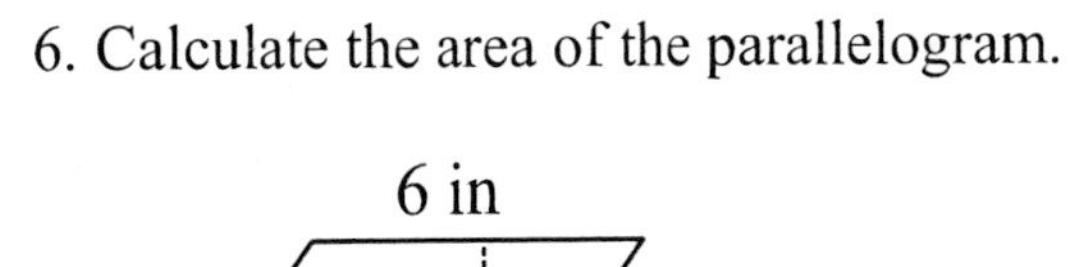

7. Calculate the area of the following figure.

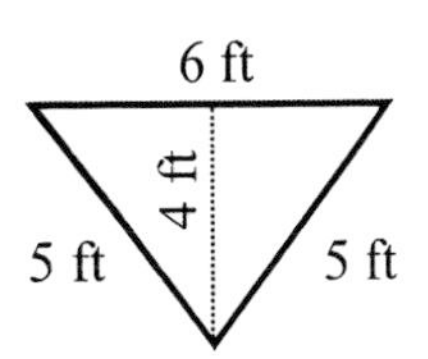

Chapter 7 Test

1. What type of triangle is illustrated?

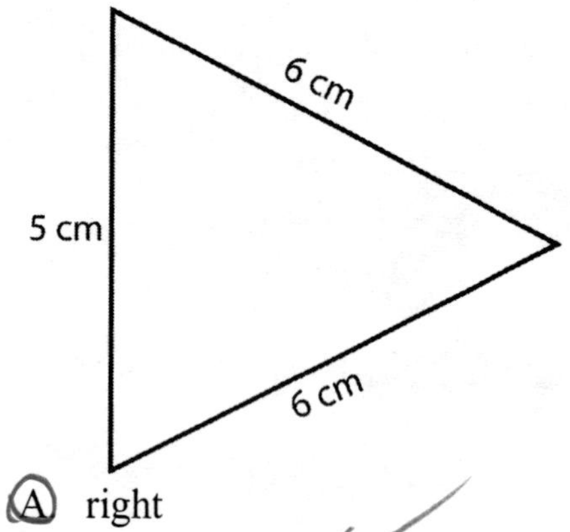

A. right
B. isosceles
C. equilateral
D. obtuse

2. Which item below is not a polygon?

A. triangle
B. heptagon
C. octagon
D. circle

3. Estimate the area.

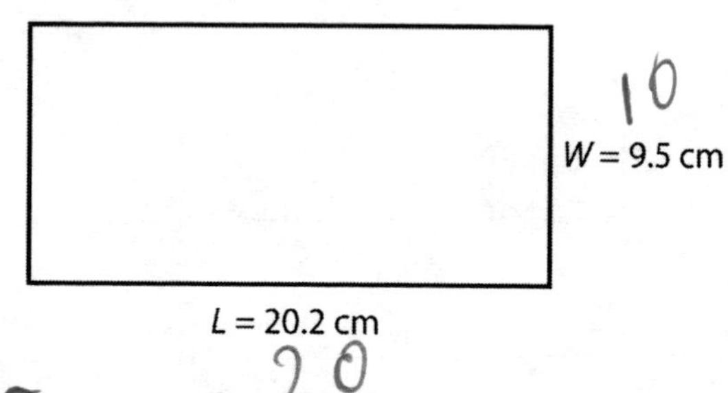

A. 200 cm^2
B. 50 cm^2
C. 30 cm^2
D. 150 cm^2

4. Find the area.

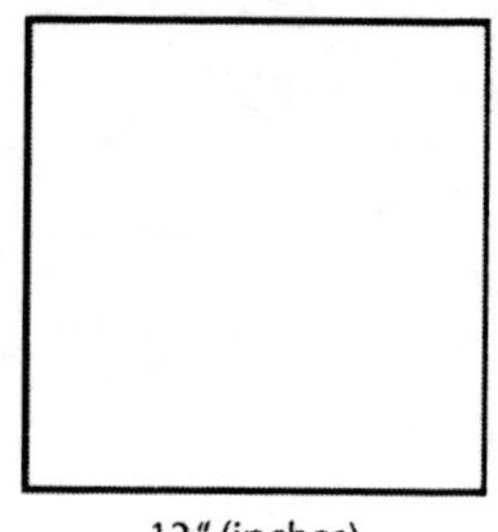

A. 48 inches^2
B. 24 inches^2
C. 132 inches^2
D. 144 inches^2

5. Find the area.

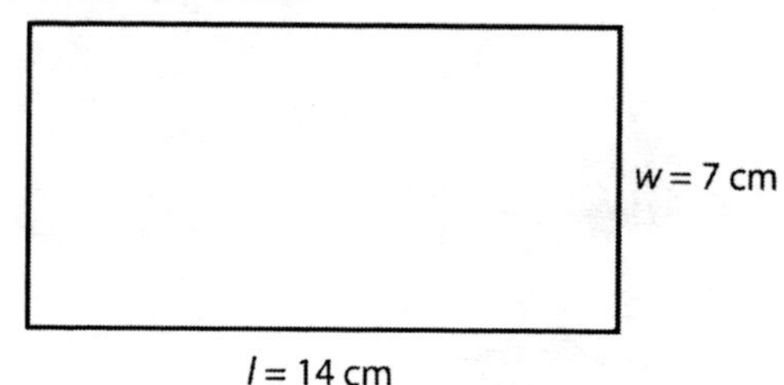

A. 98 cm^2
B. 42 cm
C. 98 cm
D. 42 cm^2

6. Find the area.

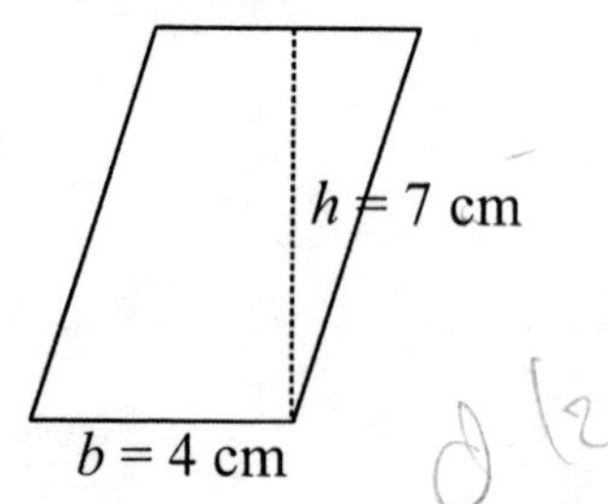

A. 28 cm
B. 28 cm^2
C. 22 cm
D. 22 cm^2

7. Find the area.

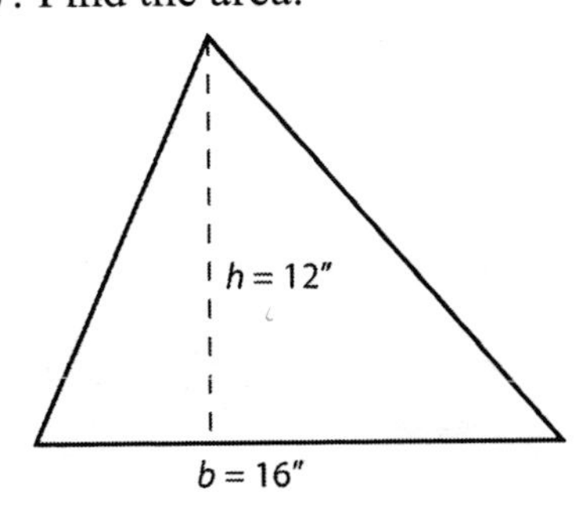

A. 96 cm
B. 28 cm^2
C. 96 cm^2
D. 28 cm

8. Find the circumference. Use $\pi = 3.14$.

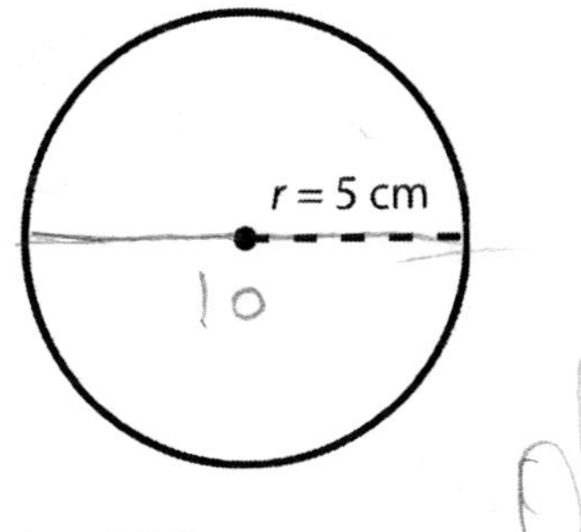

A. 15.7 cm
B. 62.8 cm
C. 31.4 cm
D. 0.314 cm

9. Find the area. Use $\pi = 3.14$.

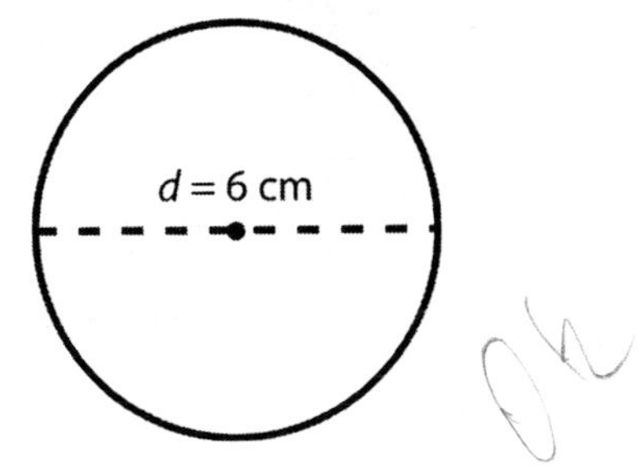

A. 113.04 cm^2
B. 28.26 cm^2
C. 18.84 cm^2
D. 188.4 cm^2

10. Ms. Hill wants to paint one wall in her living room. The dimensions are 12 feet long by 10 feet tall. One gallon of paint costs $20.00 and will cover one hundred square feet of wall space. If she purchases enough paint to cover the entire wall, how much change will she receive after paying the cashier $50?

A. $15
B. $35
C. $5
D. $10

11. Wayne is buying carpet for his bedroom. The room dimensions are 18 feet by 10 feet. The carpet cost is $3.50/square foot. How much will it cost to purchase carpet for the entire room?

A. $51.42
B. $630.50
C. $630
D. None of the above

12. Which statement is true?

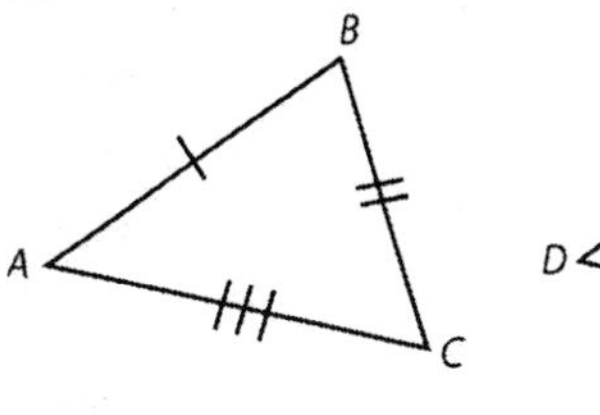

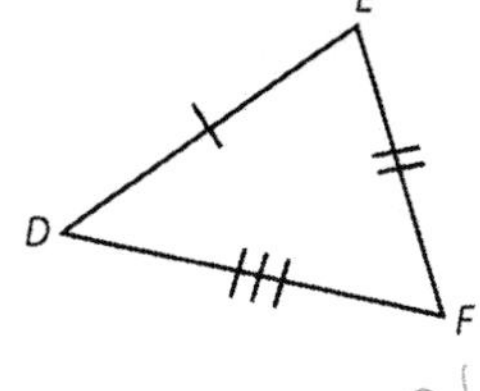

A. $\overline{AB}$ is congruent to $\overline{EF}$
B. $\overline{BC}$ is congruent to $\overline{DE}$
C. $\overline{AC}$ is congruent to $\overline{DF}$
D. $\overline{AB}$ is congruent to $\overline{DF}$

Chapter 8
Solid Geometry

This chapter covers the following Georgia Performance Standards:

M5M	Measurement	M5M4.a, b, c, d, e, f
M5P	Process Skills	M5P1.c

8.1 Understanding Volume

Volume - Measurement of volume is expressed in cubic units such as in^3, ft^3, m^3, cm^3, or yd^3. The volume of a solid is the number of cubic units that can be contained in the solid. **Capacity** is a measurement describing the greatest amount an object can hold. While volume is used to describe a 3-dimensional object's size, capacity refers to largest possible size that object could be or the largest possible amount another object could hold.

First, let's look at rectangular solids.

Example 1: How many 1 cubic centimeter cubes will it take to make up the figure below?

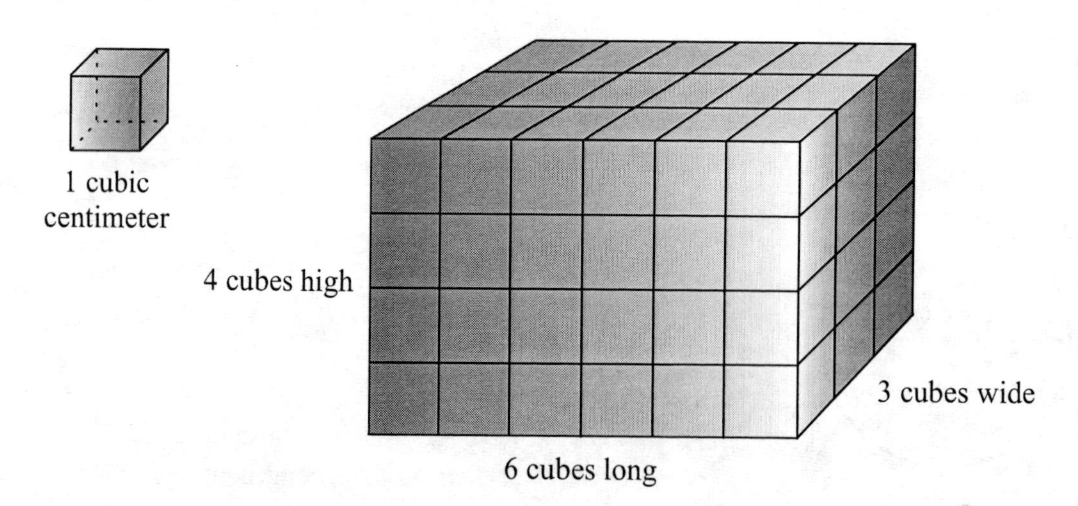

To find the volume, you need to multiply the length times the width times the height.

Volume of a rectangular solid = length × width × height $(V = lwh)$.

$$V = 6 \times 3 \times 4 = 72 \text{ cm}^3$$

Find the volume of the following rectangular solids.

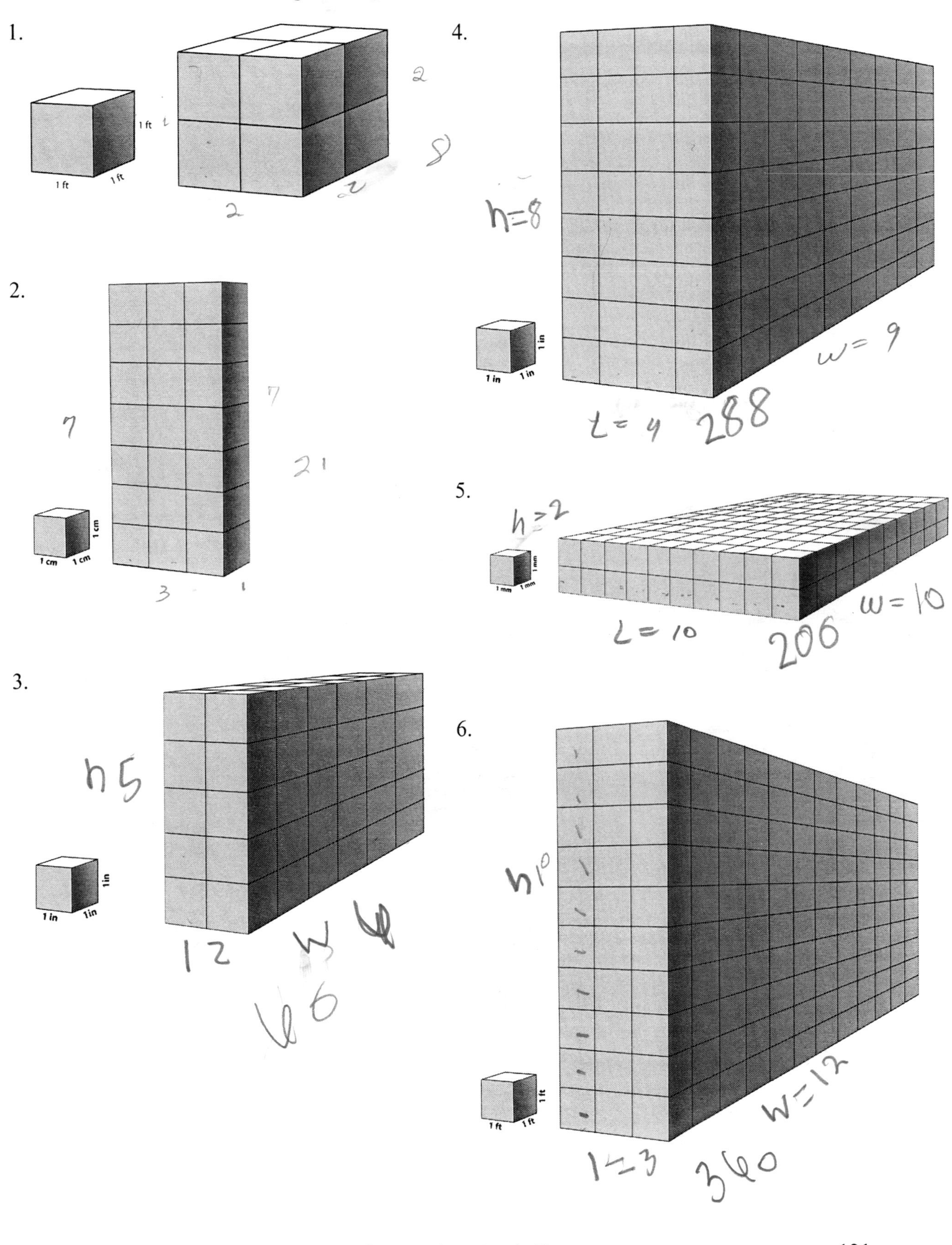

8.2 Volume of Rectangular Prisms

You can calculate the volume (V) of a rectangular prism (box) by multiplying the length (l) by the width (w) by the height (h), as expressed in the formula $V = (lwh)$.

Example 2: Find the volume of the box pictured here:

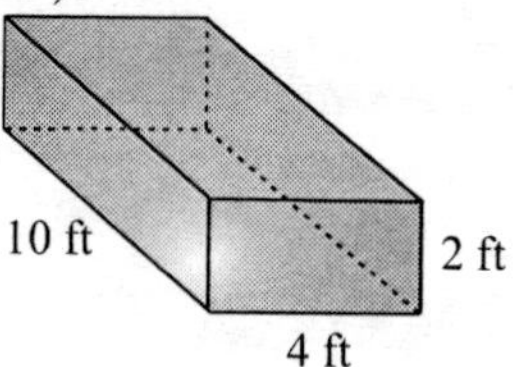

Step 1: Insert measurements from the figure into the formula.

Step 2: Multiply to solve. $10 \times 4 \times 2 = 80 \text{ ft}^3$

Note: **Volume is always expressed in cubic units such as in^3, ft^3, m^3, cm^3, or yd^3.**

Find the volume of the following rectangular prisms (boxes).

1.
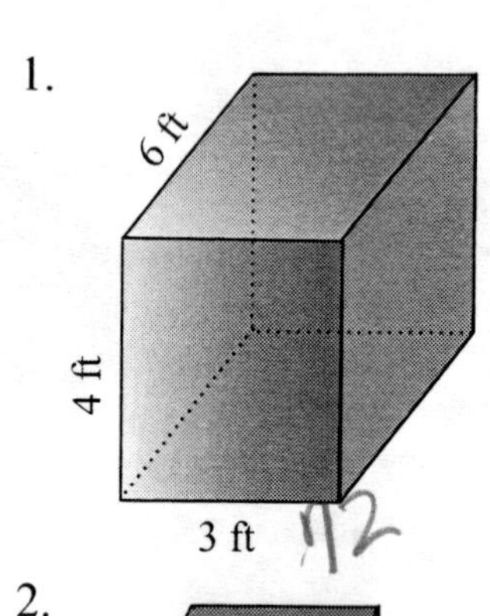

2.
10 m
15 m
8 m

3.
6 cm
8 cm
5 cm

4.
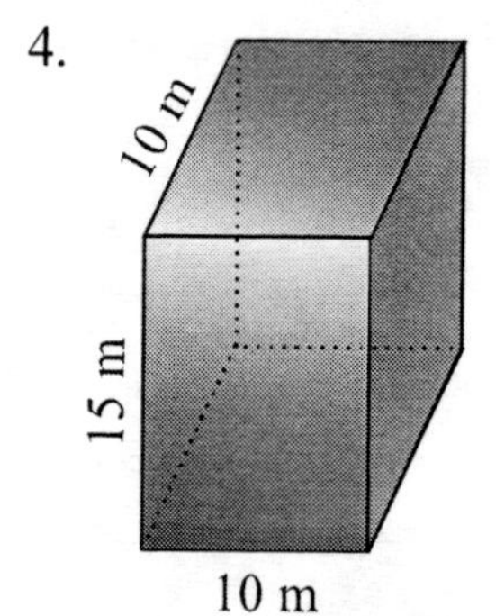

5.
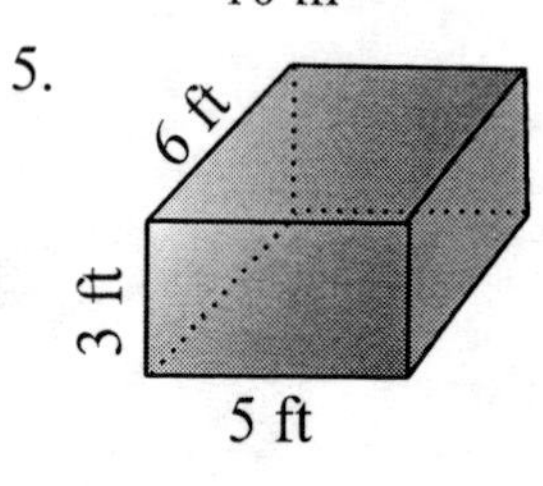

6.
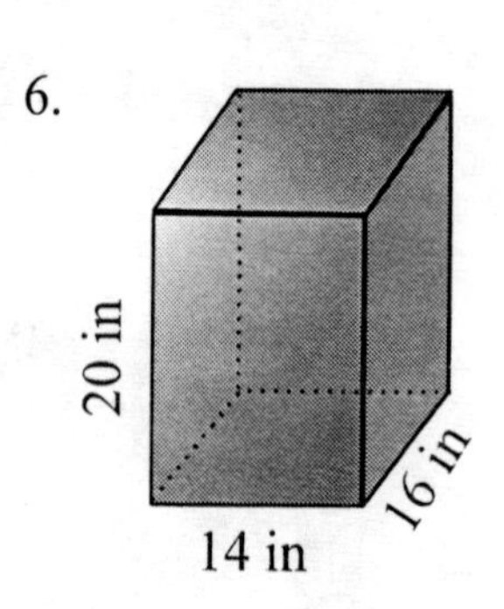

7.
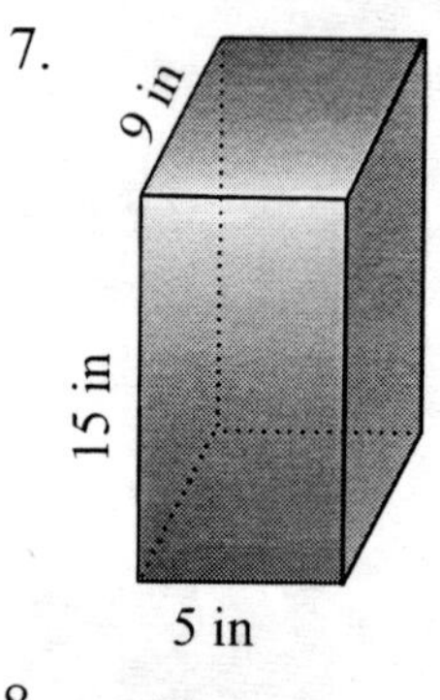

8.
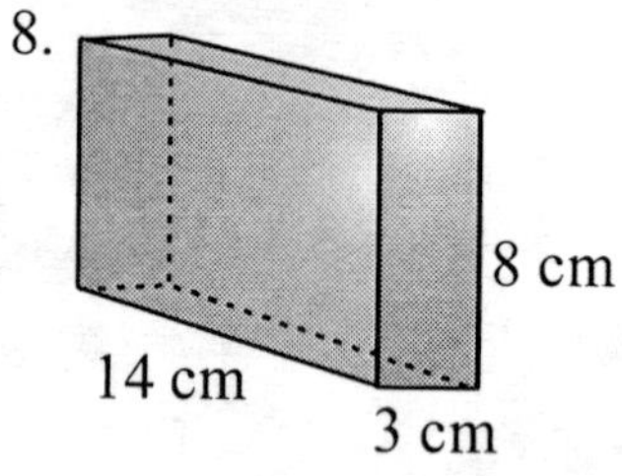

9.
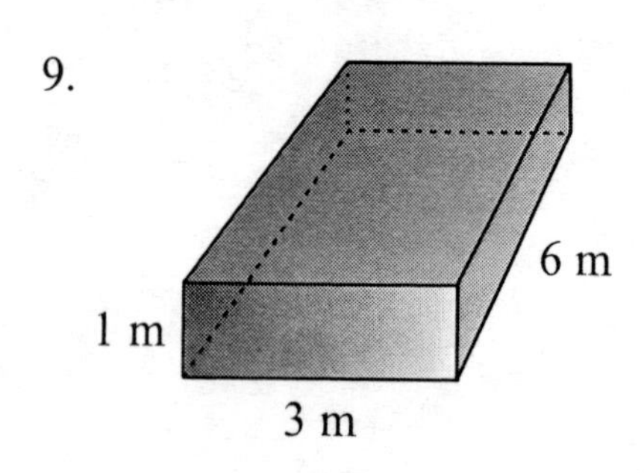

8.3 Volume of Cubes

A **cube** is a special kind of rectangular prism (box). Each side of a cube has the same measure. So, the formula for the volume of a cube is $V = s^3$ $(s \times s \times s)$.

Example 3: Find the volume of the cube at right:

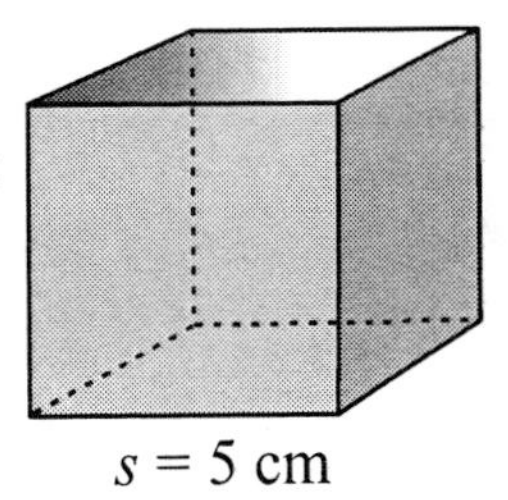

$s = 5$ cm

Step 1: Insert measurements from the figure into the formula.

Step 2: Multiply to solve. $5 \times 5 \times 5 = 125 \text{ cm}^3$

Note: Volume is always expressed in cubic units such as in^3, ft^3, m^3, cm^3, or mm^3.

Answer each of the following questions about cubes.

1. If a cube is 3 centimeters on each edge, what is the volume of the cube?

27

2. If the measure of the edge is doubled to 6 centimeters on each edge, what is the volume of the cube?

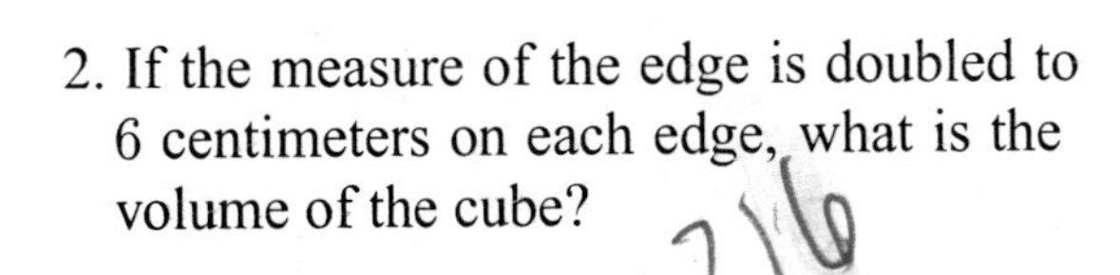

216

3. What if the edge of a 3 centimeter cube is tripled to become 9 centimeters on each edge? What will the volume be?

729

4. How many cubes with edges measuring 3 centimeters would you need to stack together to make a solid 12 centimeter cube?

64

5. What is the volume of a 2-centimeter cube?

8

6. Jerry built a 2-inch cube to hold his marble collection. He wants to build a cube with a volume 8 times larger. How much will each edge measure?

4

Find the volume of the following cubes.

7.

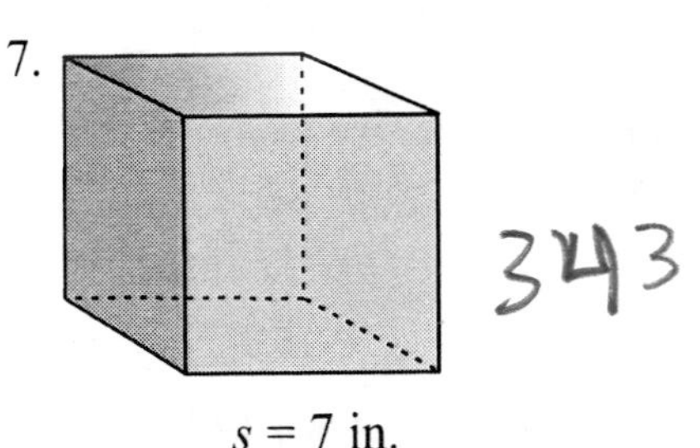

$s = 7$ in.

343

8.

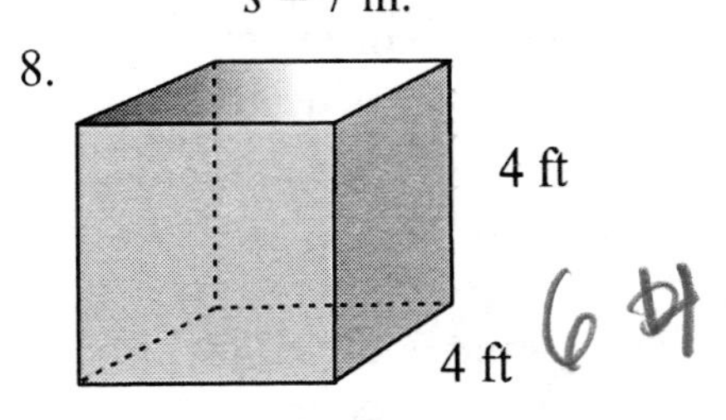

4 ft

64

9. 12 inches = 1 foot

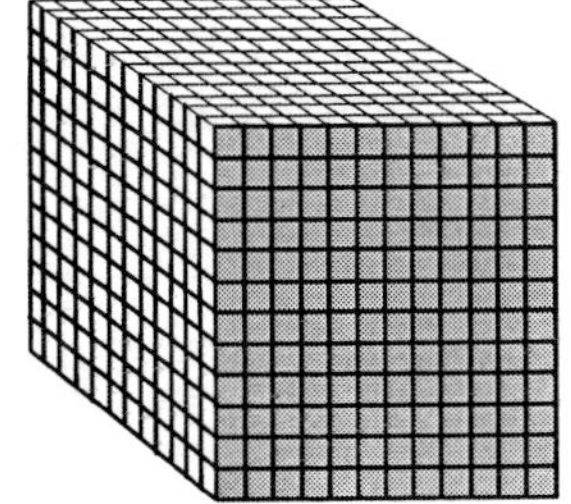

$s = 1$ foot

How many cubic inches are in a cubic foot?

1608

Chapter 8 Review

1. What is the volume of the following figure?

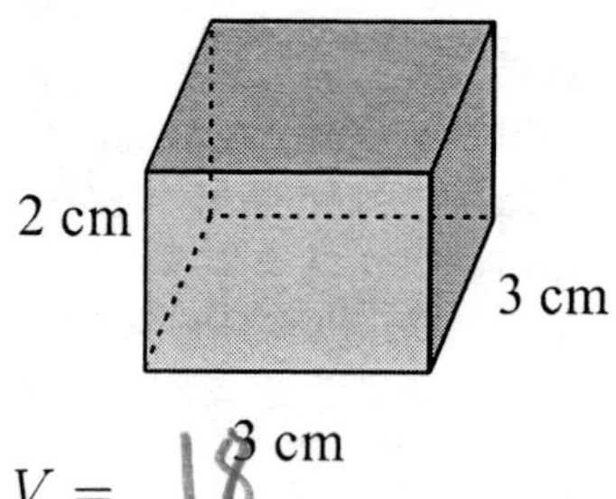

$V =$ ________

2. The sandbox at the local elementary school is 60 inches wide and 100 inches long. The sand in the box is 6 inches deep. How many cubic inches of sand are in the sandbox?

3.

2 in

2 in

2 in

It takes 8 cubic inches of water to fill the cube above. If each side of the cube is doubled, how much water is needed to fill the new cube?

4. Find the volume of the figure below. Each side of each cube measures 4 feet.

5. Find the volume of the figure below.

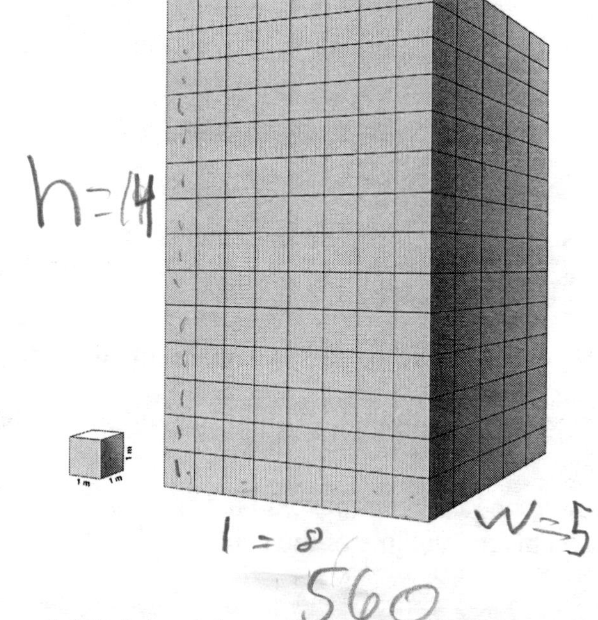

6. Find the volume of the figure below.

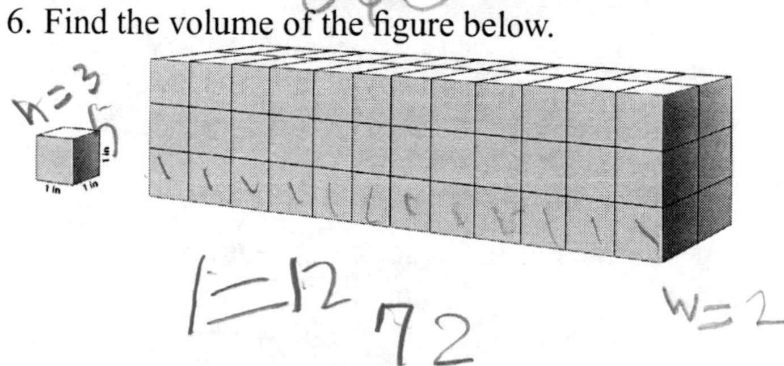

7. Find the volume of the figure below.

Chapter 8 Test

1. Find the volume.

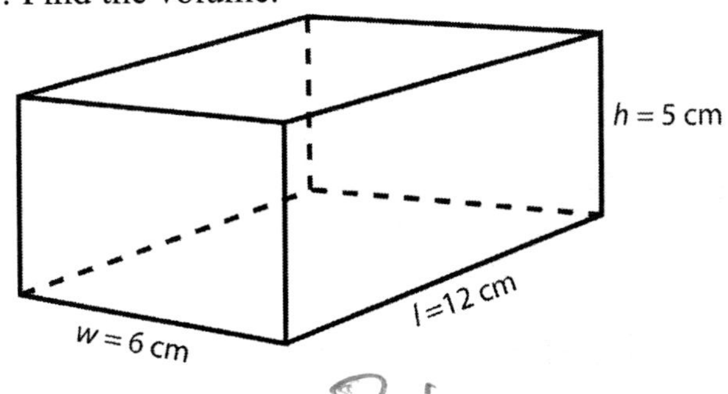

A. 72 cm^3
B. 96 cm^3
C. 360 cm^3
D. 77 cm^3

2. Find the volume of the cube.

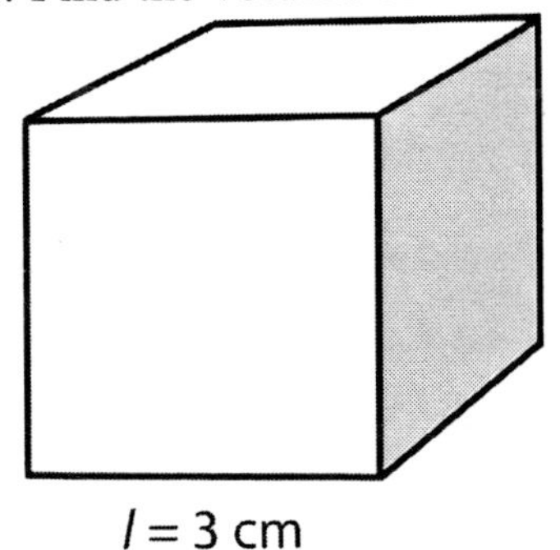

A. 36 cm^3
B. 9 cm^3
C. 27 cm^3
D. 12 cm^3

3. Find the volume of the figure below.

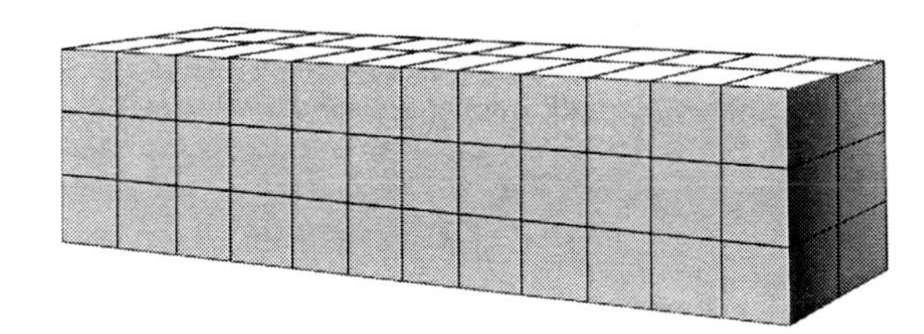

A. 24 units^3
B. 36 units^3
C. 72 units^3
D. 108 units^3

4. Find the volume.

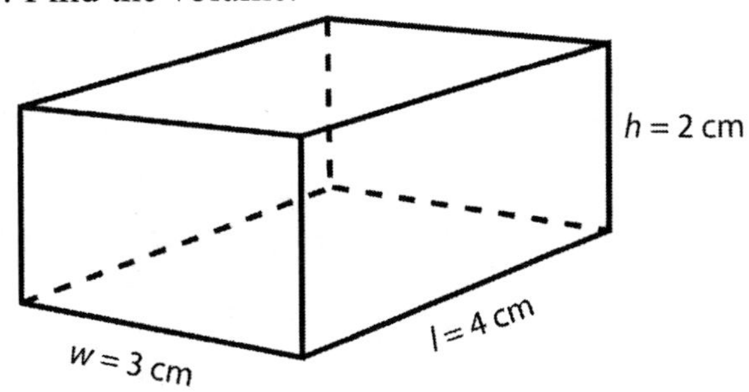

A. 24 cm^3
B. 12 cm^3
C. 36 cm^3
D. 9 cm^3

5. Find the volume of the cube.

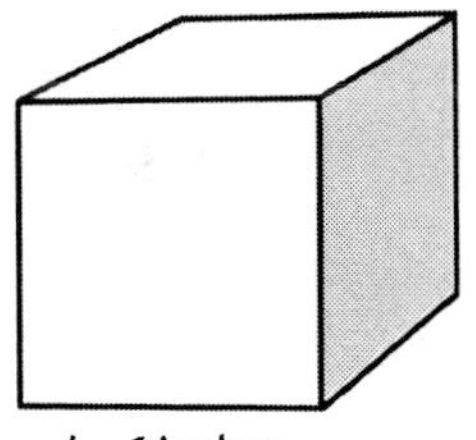

A. 36 inches^3
B. 216 inches^3
C. 864 inches^3
D. 18 inches^3

6. Find the volume of the figure below.

A. 6 units^3
B. 12 units^3
C. 36 units^3
D. 72 units^3

7. Find the volume of the figure below.

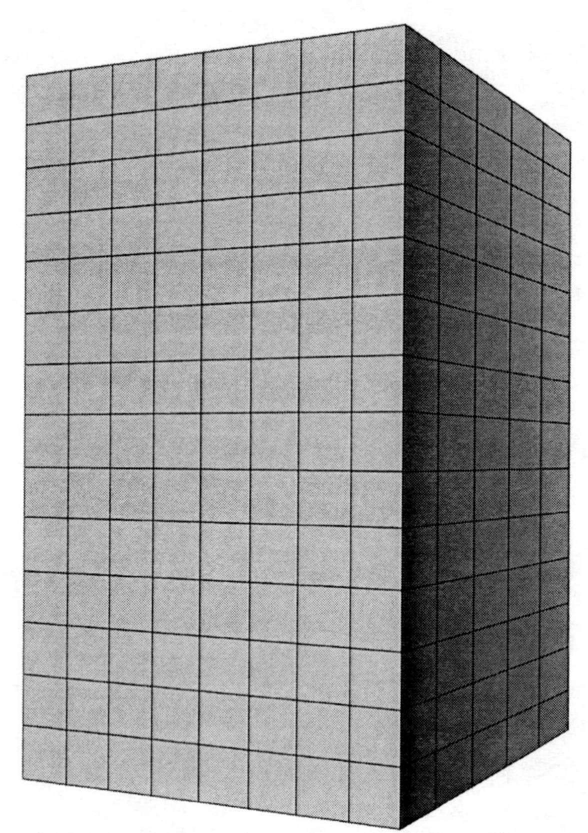

A. 672 $units^3$
B. 112 $units^3$
C. 27 $units^3$
D. 560 $units^3$

8. Find the volume of the figure below.

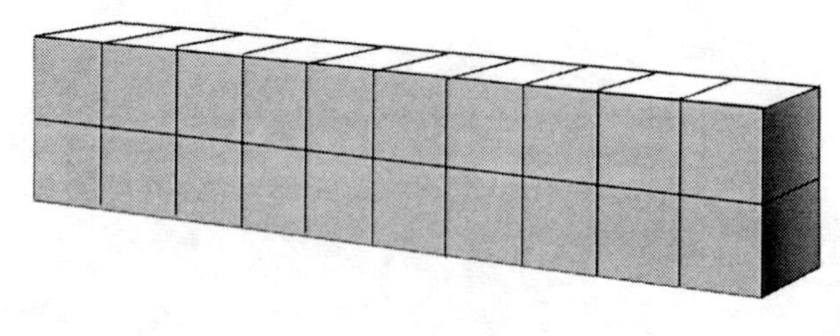

A. 20 $units^3$
B. 21 $units^3$
C. 120 $units^3$
D. 10 $units^3$

9. Find the volume of the figure below.

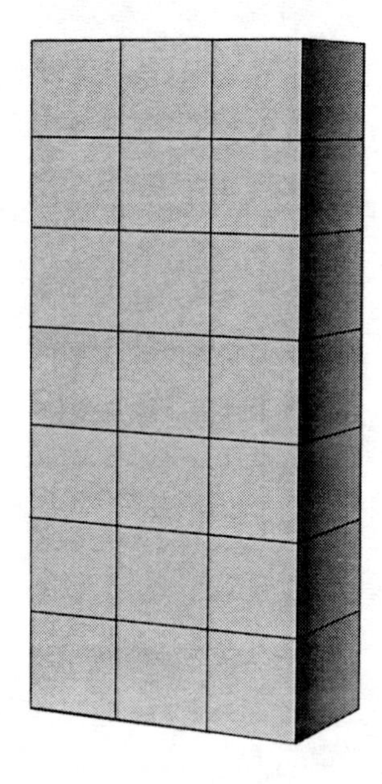

A. 22 $units^3$
B. 7 $units^3$
C. 21 $units^3$
D. 11 $units^3$

10. Find the volume of the figure below.

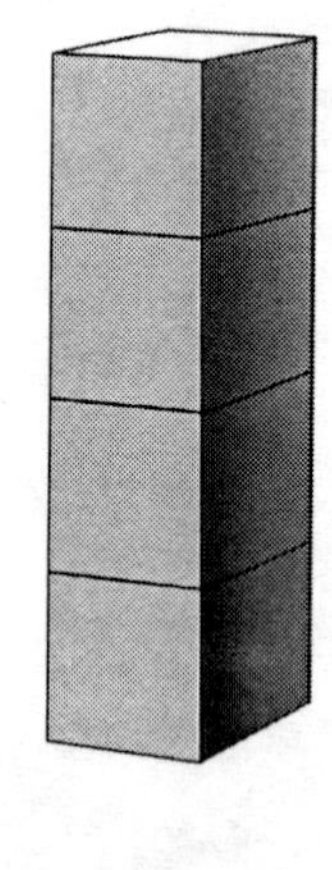

A. 1 $unit^3$
B. 4 $units^3$
C. 6 $units^3$
D. 12 $units^3$

Chapter 9
Data Interpretation

This chapter covers the following Georgia Performance Standards:

M5N	Numbers and Operations	M5N5.b
M5D	Data Analysis	M5D1.a, b M5D2
M5P	Process Skills	M5P1.a, b, c M5P5.a, b, c

9.1 Reading Tables

A **table** is a concise way to organize large quantities of information using rows and columns.

Example 1: Students at Oak View Elementary were asked which flavor of ice cream was their favorite. How many 5th graders preferred mint chocolate ice cream?

	Type of Ice Cream			
Grade	Vanilla	Chocolate	Mint Chocolate	Other
K-3	379	284	231	48
4	69	81	41	27
5	99	64	39	11
6	49	40	21	17

Step 1: Take a look at the grade intervals on the left column. The 5th grade is the row we will be looking at.

Step 2: Now look at the types of ice cream on the top row. We want to find the box labeled mint chocolate. This is the column we will be looking at.

Step 3: Using the row and column we found in steps 1 and 2, locate the box where the row and column intersect. They intersect at 39. The number of 5th graders whose favorite ice cream is mint chocolate is 39.

Jimmy wants to know how many girls and boys, aged 12, play either football; basketball; or soccer. He goes to the park after school and surveys 80 kids. The table below shows his results. Read each table carefully, and then answer the questions that follow.

	Type of Sport		
	Football	Basketball	Soccer
Boys	14	22	16
Girls	1	15	12

1. How many girls play soccer?

2. How many boys play football or basketball?

3. How many boys and girls play basketball?

4. How many boys play soccer?

5. How many girls play basketball?

6. How many boys and girls play soccer?

9.2 Bar Graphs

Bar graphs can be either vertical or horizontal. There may be just one bar or more than one bar for each interval. Sometimes each bar is divided into two or more parts. The data is made up of more than one category. In this section, you will work with a variety of bar graphs. Be sure to read all titles, keys, and labels to completely understand all the data that is presented.

Example 2: The results for last year's high school math grades have been calculated. How many more 9th graders made A's in math than 11th graders?

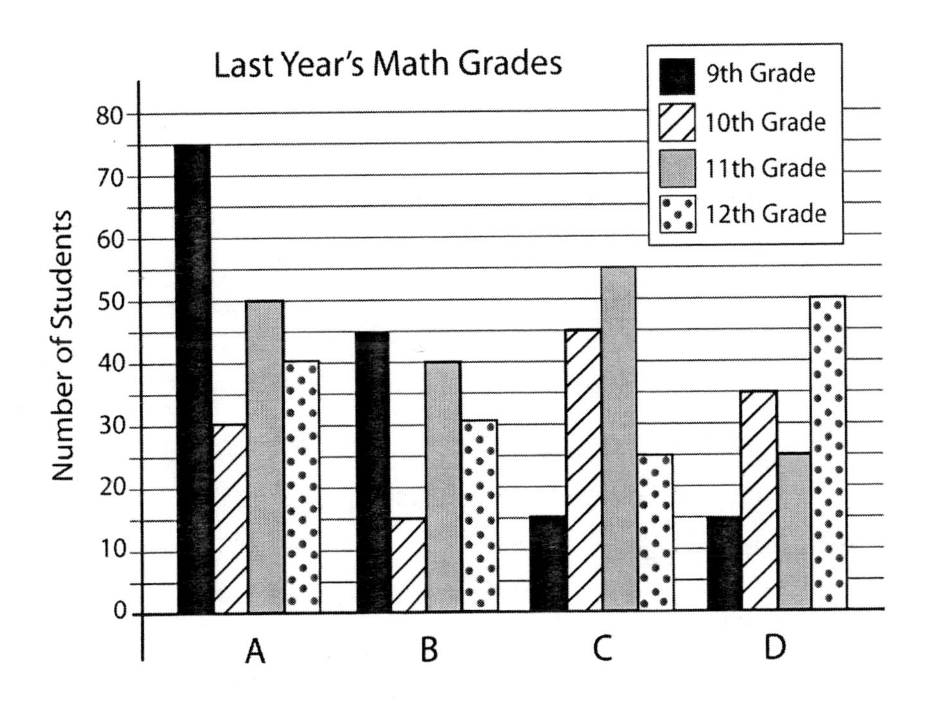

Step 1: Looking under the category of A's, locate the graph for 9th graders. The interval on the left side of the bar graph reads 75 at the highest point on the graph for 9th graders. That means 75 9th graders made an A in math class.

Step 2: Now do the same thing in step 1 again except this time instead of looking at the graph for 9th graders, we look at the graph for 11th graders. The interval on the left side of the bar graph reads 50 at the highest point on the graph for 11th graders. That means 50 11th graders made an A in math class.

Step 3: Subtract the 50 11th graders from the 75 9th graders. There is an excess of 25 students. This means that 25 more 9th graders made an A in math than 11th graders.

Answer the questions about each graph.

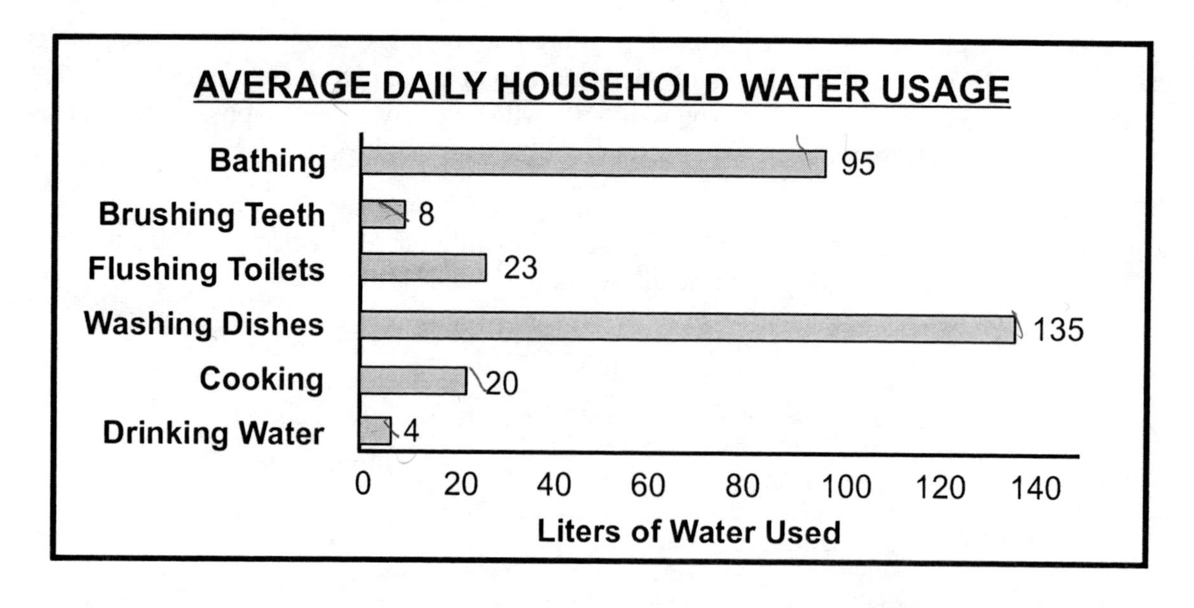

1. How many liters of water does the average household use every day by washing dishes and cooking?
2. How many liters of water would an average household drink in a week (7 days)?
3. How many liters of water does an average household use in a day?
4. How many more liters does an average household use for bathing than it does for cooking?
5. How many liters does the average household use every day to bathe and brush teeth?

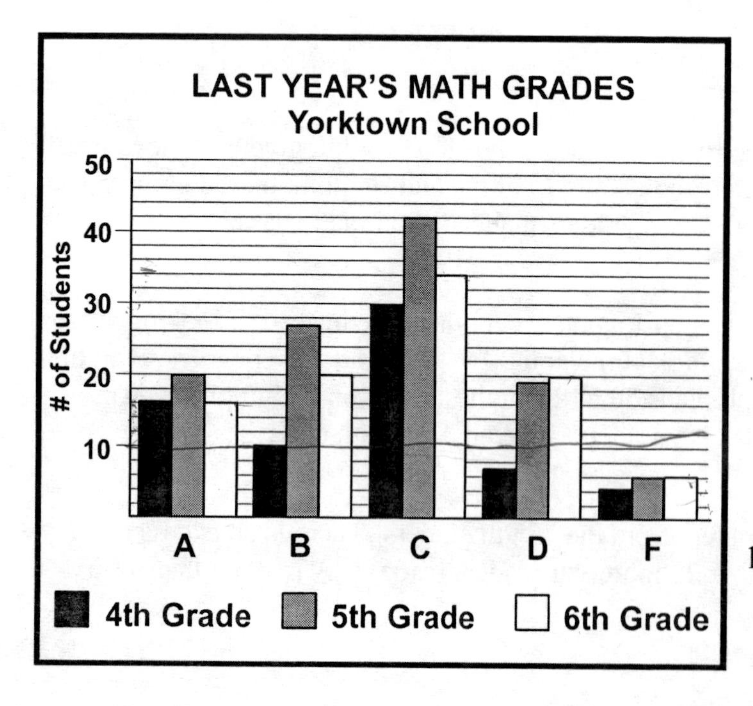

6. How many of last year's 4th graders made C's in math?
7. How many more math students made B's in the 5th grade than in the 6th grade?
8. Which letter grade occurs the most number of times in the data?
9. How many 6th graders took math last year?
10. How many students made A's in math last year?

9.3 Line Graphs

Line graphs often show how data changes over time. Time is always shown on the bottom of the graph. When choosing a graph to show how something changes over time, a line graph is the best choice.

Example 3: Pete is comparing how well his peanut farm has done in recent years. How many more peanuts were harvested in the 2002 than were harvested in 1999?

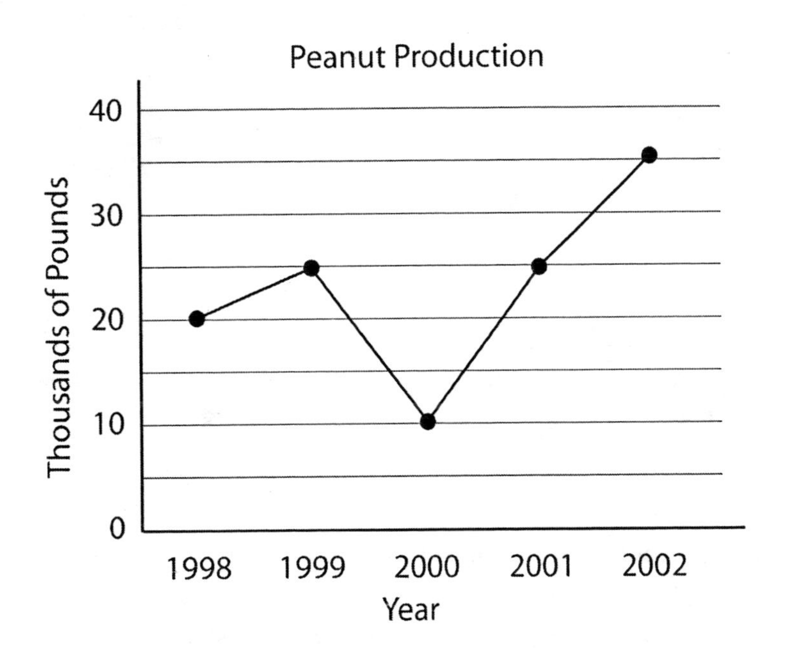

Step 1: Locate the year 2002 from the row of years at the bottom of the line graph. Find the dot on the graph above 2002. Then follow the line to the left to see how many thousands of pounds of peanuts were sold. It is half way between 30 and 40. Halfway between 30 and 40 is 35. From the label on the side of the graph, we know this means thousands of pounds. The graph shows 35,000 pounds of peanuts produced in 2002.

Step 2: Read the graph for the year 1999. Now locate the year 1999 from the row of years at the bottom of the line graph. The graph shows that 25,000 lbs of peanuts were produced in 1999.

Step 3: Subtract 35,000 lbs from 25,000 lbs, and you get 10,000 lbs. In 2002, 10,000 more pounds of peanuts were produced than in 1999.

Study the line graphs below, and then answer the questions that follow.

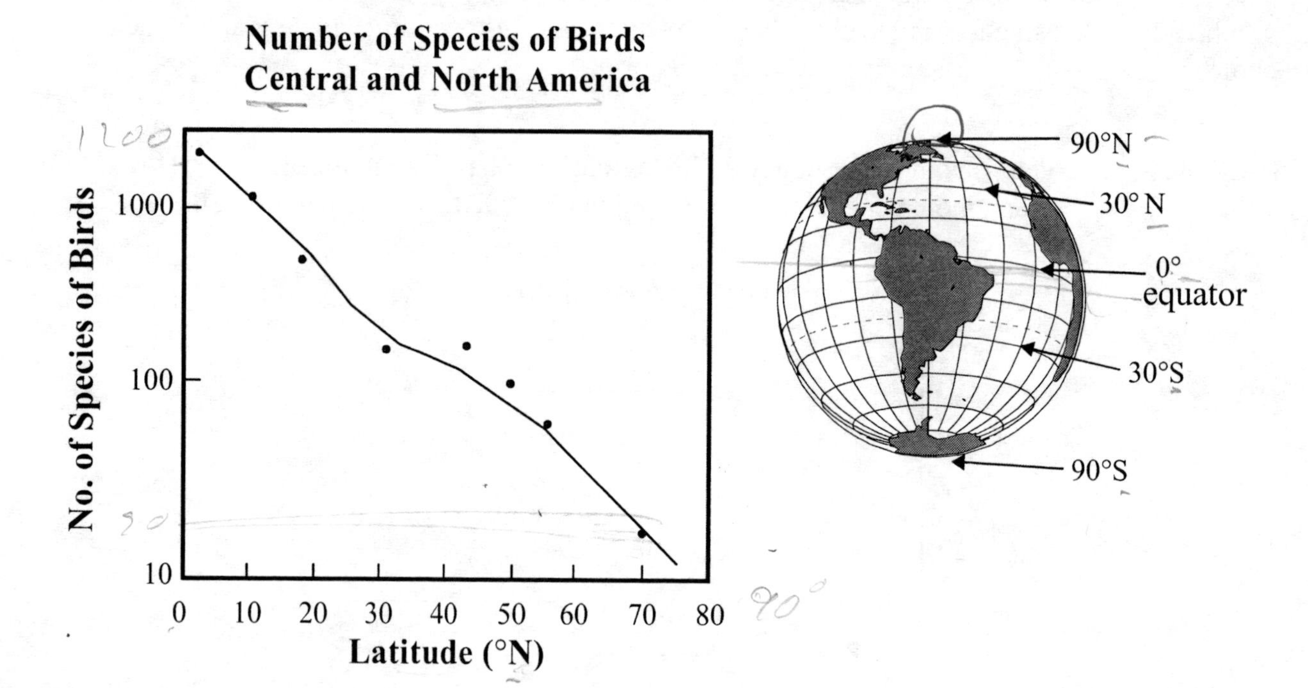

After reading the graph above, label each of the following statements as true or false.

1. Where are the most species of birds?

   A. The North Pole
   B. the equator
   C. 30°N
   D. 60°N

2. Which state would most likely have the fewest species of birds?

   A. Georgia
   B. Alabama
   C. Florida
   D. Alaska

3. About how many species of birds are at 30°N according to the graph?

   A. 30
   B. 60
   C. 90
   D. 110

4. At 90°N, we find the North Pole, about how many different species of birds can we expect to find there based on the graph?

   A. 1
   B. 10
   C. 100
   D. 1,000

9.4 Circle Graphs

Circle graphs represent data expressed in percentages of a total. The parts in a circle graph should always add up to 100%. Circle graphs are sometimes called **pie graphs** or **pie charts**. A few examples of circle graphs are: percentage of kids who prefer a peanut butter and jelly sandwich; or the least favorite color between red, yellow, and purple.

Example 4: A survey of 1,000 students was taken to see which flavor of ice cream should be added to the list of flavors in the school cafeteria: mint chocolate chip, strawberry, vanilla, banana, or pecan cluster. How many students voted for strawberry?

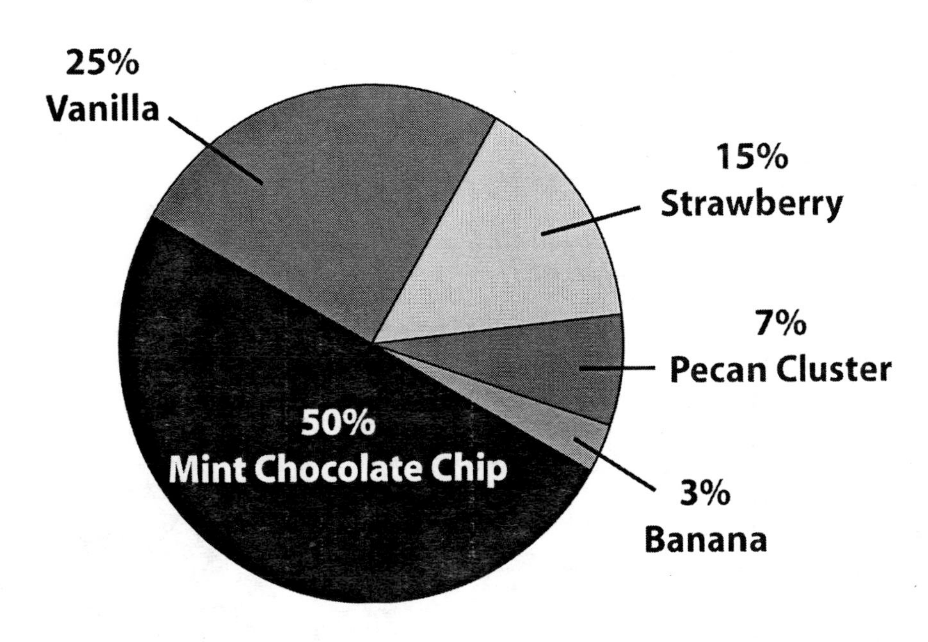

Step 1: Locate the category "Strawberry" on the circle graph. 15% of students choose strawberry.

Step 2: Multiply the percentage of students whom chose strawberry to the total number of students. $0.15 \times 1,000 = 150$. Of the 1,000 students surveyed, 150 of them chose voted for strawberry as their favorite flavor of ice cream.

To figure the value of a percent in a circle graph, multiply the percent by the total. Use the circle graphs below to answer questions. The first question is worked for you as an example.

1. How much does Tina spend each month on music CD's?

   $80 × 0.20 = $16.00 **$16.00**

2. How much does Tina spend each month on make-up? ______

3. How much does Tina spend each month on clothes? ______

4. How much does Tina spend each month on snacks? ______

Fill in the following chart.

Favorite Activity	Number of Students
5. watching TV	
6. talking on the phone	
7. playing video games	
8. surfing the Internet	
9. playing sports	
10. reading	

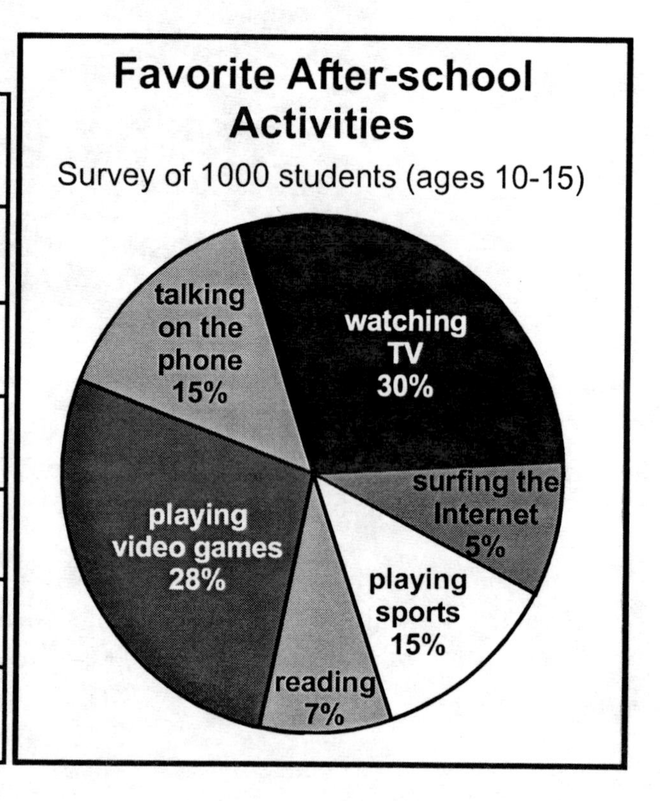

9.5 Comparing Types of Graphs

Example 5: Sandy is measuring the temperature outside her house once an hour on the hour from 8:00 in the morning to 8:00 at night. After she gathers all the data, what type of graph should she make to illustrate the data?

Step 1: Find the variable that is being measured. We want to know the different temperatures outside, so temperature is being measured.

Step 2: Find out how often temperature is being measured. There will be 12 measurements taken, one every hour for 12 hours.

Step 3: Each new measurement is a continuation of the last. Which type of graph has continuous data? Line graphs use continuous data. Sandy should use a line graph to illustrate her data.

For questions 1–5 answer either: line graph, circle graph, or bar graph.

1. Tucker Enterprises is conducting a study of 2,000 families in the area to find out what percent of families have 1 child, 2 children, 3 children, or 4 or more children. The data has been gathered. What type of graph would be most appropriate to show the results of the study?
2. Ellen is performing a science project for school. She wants to see if age plays a role in the time it takes a person's heart rate to return to normal after 30 minutes of exercise. As part of her results, she must graph the data for exercise and heart rate. Which type of graph should she use?
3. Which type of graph would be most appropriate to show the population changes for the state of Texas from 1997 to 2001?
4. Mrs. Reed's class sold the most candy bars for the fund-raiser. As a reward, the class gets to watch a movie on Friday. The class voted between five different movies to watch. What type of graph should Mrs. Reed make to post the results?
5. Sales are down at Joe's Ice Cream Shop. Joe has decided he needs a new fresh flavor to boost sales. He has picked three possible new flavors, but he wants his customers to make the final decision. All week ballot sheets have been in the shop for customers to vote. What type of graph should Joe use to show the results?

For questions 6–7 answer the question.

6. Sammy wants to make a graph that shows how much time he spends on each main task he has to do throughout the day. He decides to make a bar graph. Is this the correct type of graph Sammy should make?
7. Jim is helping Katie with an after-school project. She is supposed to graph the population for each state in 2003. Katie thinks she should make a line graph. Jim thinks she should make a bar graph. Who is right?

Chapter 9 Review

KNIGHTS BASKETBALL Points Scored				
Player	**game 1**	**game 2**	**game 3**	**game 4**
Joey	5	2	4	8
Jason	10	8	10	12
Brandon	2	6	5	6
Ned	1	3	6	2
Austin	0	4	7	8
David	7	2	9	4
Zac	8	6	7	4

1. How many points do the Knights basketball team score in game 1?

2. How many more points does David score in game 3 than in game 1?

3. How many points does Jason score in the first 4 games?

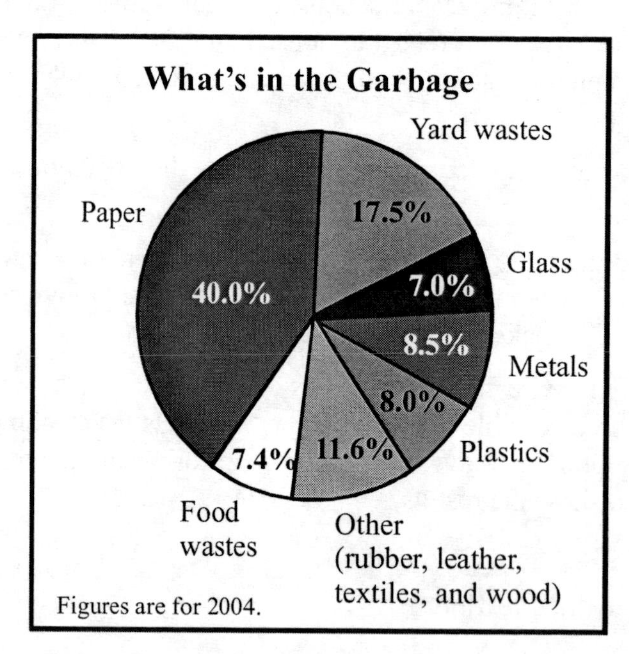

4. In 2004, the United States produced 160 million metric tons of garbage. According to the pie chart, how much glass was in the garbage?

5. Out of the 160 million metric tons of garbage, how much was glass, plastic, and metal?

6. If in 2006, the garbage reaches 200 million metric tons, and the percentage of wastes remains the same as in 2004, how much food will be in the 2006 garbage?

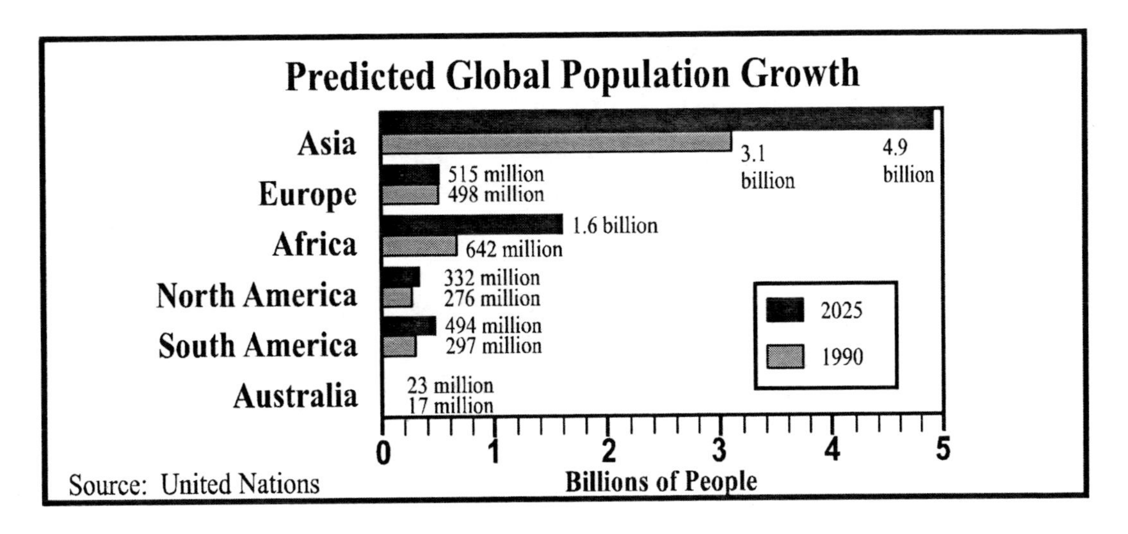

7. By how many is Asia's population predicted to increase between 1990 and 2025?

8. In 1990, how much larger was Africa's population than Europe's?

9. Where is the population expected to more than double between 1990 and 2025?

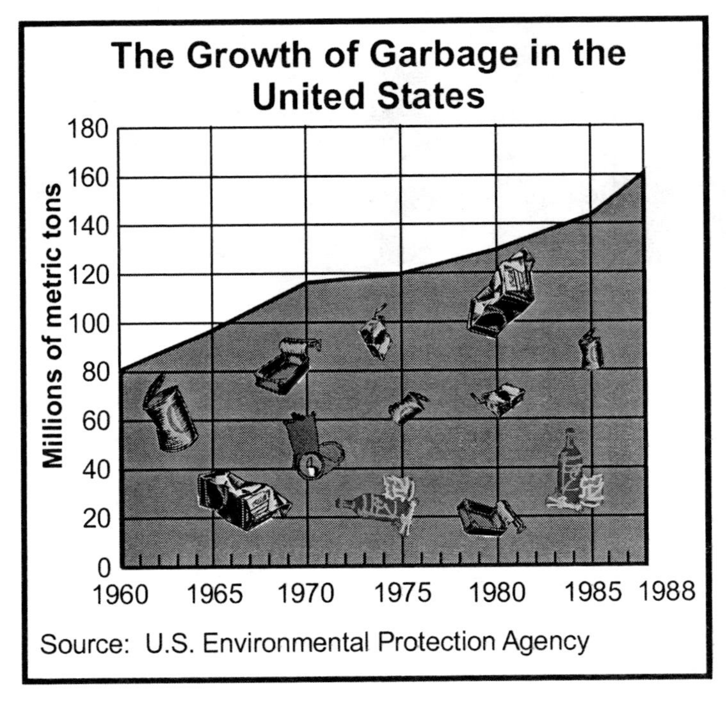

10. How much did the volume of garbage grow between 1960 and 1988?

11. In which year did garbage in the United States reach 140 million metric tons?

12. How much did the volume of garbage grow between 1960 and 1966?

Chapter 9 Test

Read the table below, and answer questions 1 and 2.

Name	Total CDs owned
Maggie	97
Erica	164
John	81
Philip	151
Tanya	122

1. Which person has about twice as many CDs as John?

   A. Philip
   B. Tanya
   C. Erica
   D. Maggie

2. Which person owns about 20% less than Philip?

   A. Maggie
   B. Tanya
   C. John
   D. Erica

Use the bar graph below, and answer questions 3 and 4.

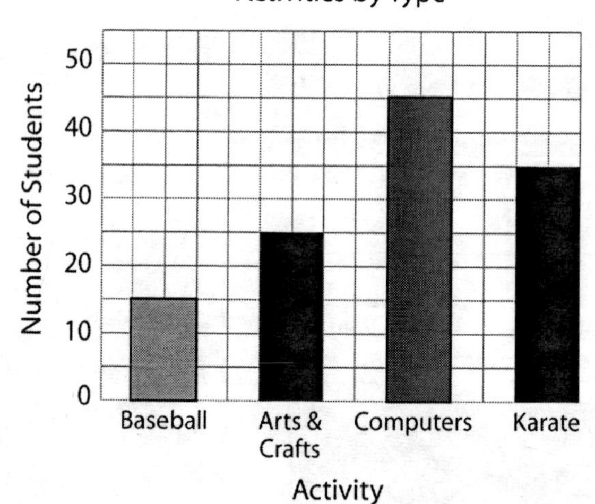

3. Which after school activity was the most popular?

   A. Basketball
   B. Arts & Crafts
   C. Computers
   D. Karate

4. The After-School Coordinator anticipates twice as many participants in karate next year. Each class has a maximum of 20 students per class. If the coordinator is correct, how many classes will she need?

   A. 3
   B. 1
   C. 4
   D. 5

Use the line graph below, and answer questions 5 and 6.

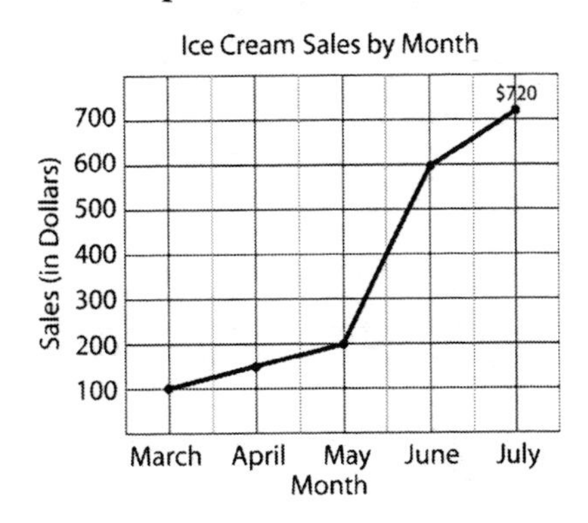

5. Which two months marked the greatest increase in sales?

   A. May – June
   B. June – July
   C. March – April
   D. April – May

6. If the owner expects sales in August to be 10% higher than July, how much should sales be in August?

   A. $902
   B. $750
   C. $792
   D. $800

Use the circle graph below, and answer questions 7 and 8.

7. If 500 students were surveyed, how many students prefer cake as a favorite snack?

   A. 450 students
   B. 130 students
   C. 150 students
   D. 175 students

8. How many students prefer candy and pudding?

   A. 65 students
   B. 60 students
   C. 125 students
   D. 100 students

Formula Sheet

Circumference	Circle	$C = 2\pi r$ or $C = \pi d$
Area	Rectangle	$A = lw$ or $A = bh$
	Parallelogram	$A = bh$
	Triangle	$A = \frac{1}{2} bh$ or $A = \frac{bh}{2}$
	Circle	$A = \pi r^2$
Volume	Cube	$V = 6s$
	Rectangular Prism	$V = Bh$*
	** B represents the area of the Base of a solid figure.*	
Pi	π	$\pi \approx 3.14$
Converting Units of Measure		
		Abbreviations
Volume	1 gallon = 4 quarts	gallon = gal
	1 quart = 2 pints	quart = qt
	1 pint = 2 cups	pint = pt
	1 cup = 8 ounces	ounce = oz
Length	1 mile = 5,280 feet	mile = mi
	1 yard = 3 feet	yard = yd
	1 foot = 12 inches	foot = ft
		inches = in
Weight	16 ounces = 1 pound	pound = lb

Practice Test 1

1. Which of the following is a composite number?

   A. 19
   B. 17
   C. 16
   D. 11

   M5N1a

2. Which fraction represents the shaded area?

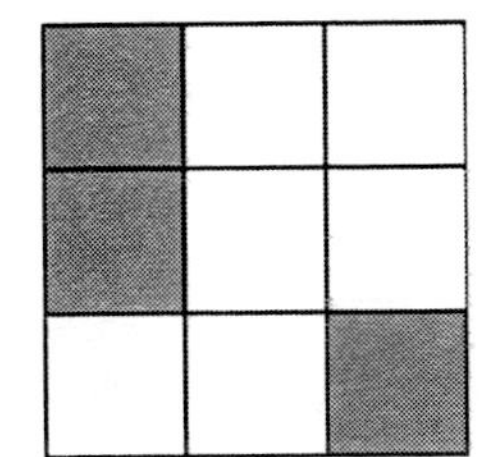

   A. $\frac{2}{3}$
   B. $\frac{3}{6}$
   C. $\frac{3}{9}$
   D. 6

   M5N4e

3. Which of the following is a prime number?

   A. 15
   B. 13
   C. 25
   D. 39

   M5N1a

4. I am a prime number less than 2. If I am multiplied by 5, my product will be ____.

   A. 5
   B. 10
   C. 15
   D. none of the above

   M5N1a

5. What percentage represents the shaded area?

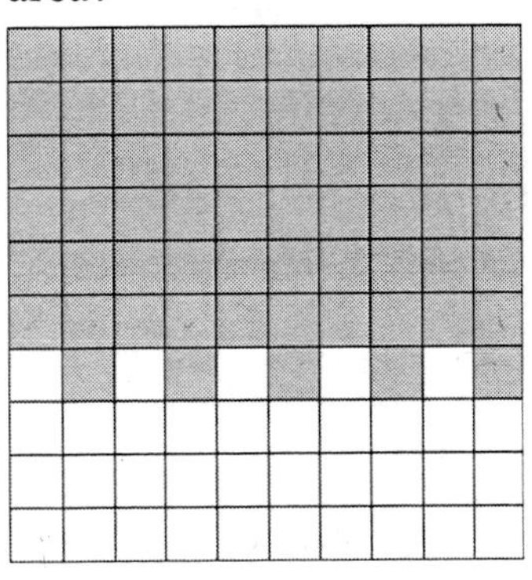

   A. 60%
   B. 62%
   C. 65%
   D. 6.5%

   M5N5a

6. Which number is a multiple of 12?

   A. 22
   B. 4
   C. 3
   D. 144

   M5N1b

7. Which number is a factor of 52?

   A. 6
   B. 14
   C. 12
   D. 13

   M5N1b

8. Which number is **not** divisible by 7?

   A. 84
   B. 49
   C. 87
   D. 28

   M5N1c

9. Estimate the area.

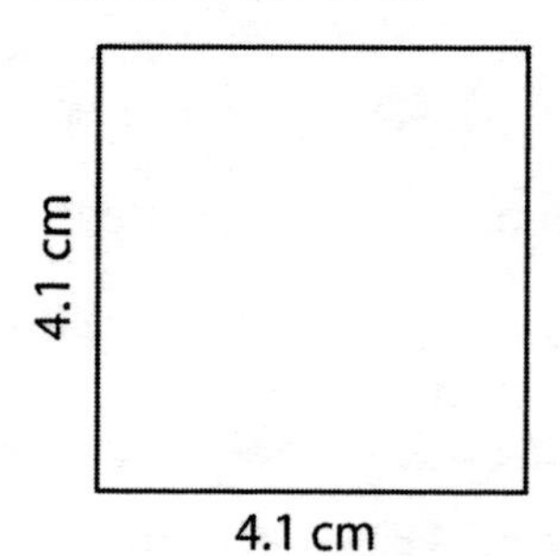

A. 12.3 cm^2
B. 16 cm^2
C. 8.2 cm
D. 16.81 cm

M5M1a

10. How many students selected the drums?

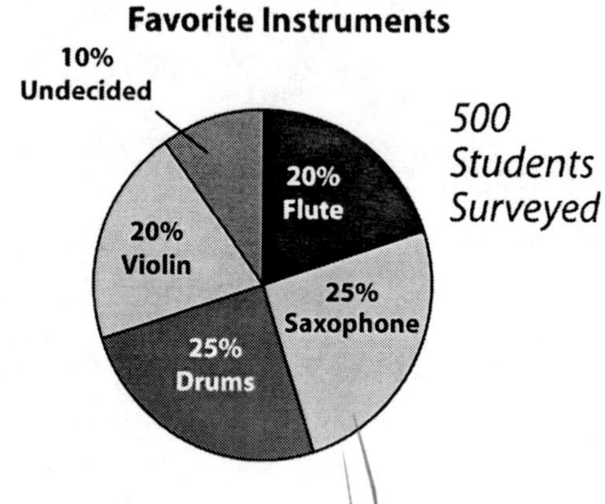

A. 100
B. 120
C. 130
D. 125

M5N5b

11. 914,005 is divisible by which number?

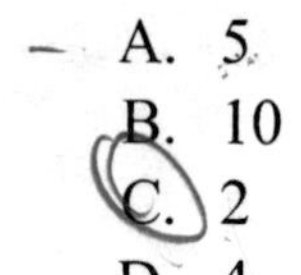

A. 5
B. 10
C. 2
D. 4

M5N1c

12. What is the place value of the underlined digit?

1<u>2</u>,067,132

A. two million
B. twenty million
C. two-hundred thousand
D. two-hundred million

M5N2a

13. Find the area.

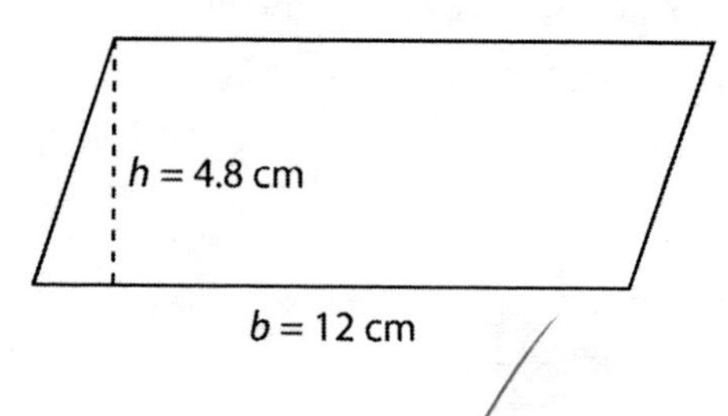

A. 48 cm^2
B. 49.6 cm^2
C. 48 cm
D. 57.6 cm^2

M5M1d

14. What is the place value of the underlined digit?

154.87<u>6</u>

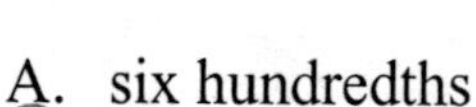

A. six hundredths
B. six thousandths
C. six tenths
D. six hundred

M5N2a

15. Find the product.

$6,228 \times 1,000$

A. 6,228,000
B. 622,800
C. 62,280
D. 602,280

M5N2b

16. Find the area.

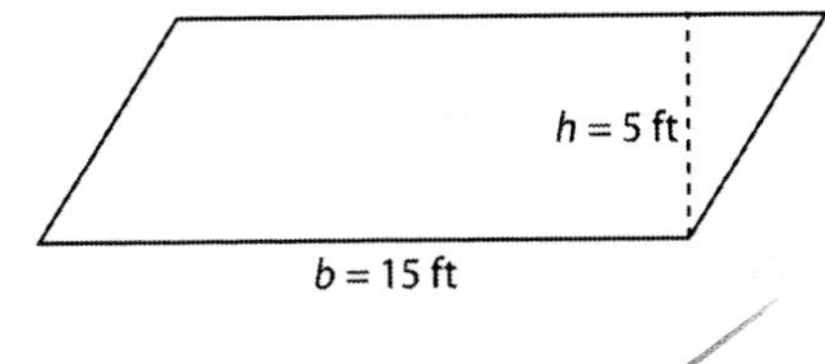

A. 40 ft^2
B. 75 ft^2
C. 37.5 ft^2
D. 65 ft^2

M5M1d

17. Find the product.

$25,690 \times 0.1$

A. 2,569
B. 256.9
C. 25.69
D. 2.569

M5N2b

18. Find the area.

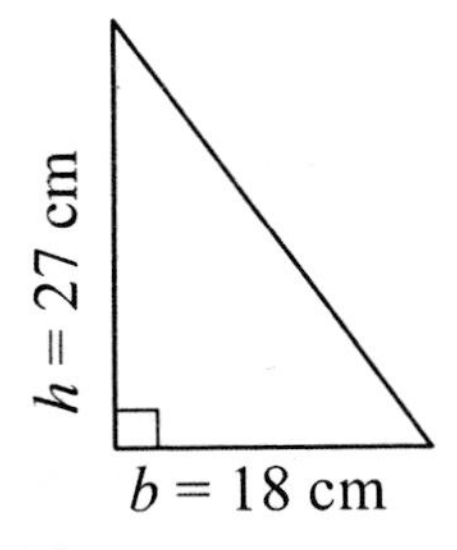

A. 486 cm^2
B. 243 cm
C. 486 cm
D. 243 cm^2

M5M1d

19. Find the product.

$13,820 \times 0.01 =$ ______

A. 1,382
B. 138.2
C. 138,200
D. 13.82

M5N2b

20. $682.1 \times 3.1 =$ ______

A. 212
B. 211.45
C. 2114.51
D. 210.4

M5N3a

21. Which group shows numbers ordered from least to greatest?

A. 0.002, 0.0002, 0.02
B. 0.135, 1.035, 0.1355
C. 0.127, 0.0127, 0.00127
D. 0.0067, 0.067, 0.607

M5N2a

22. Find the area of the regular polygon illustrated below.

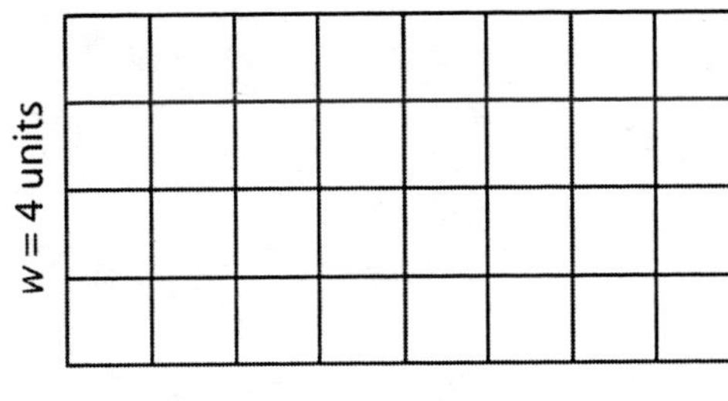

A. 24 sq. units
B. 12 sq. units
C. 32 sq. units
D. 64 sq. units

M5M1f

23. 7500 mL = ________ L

A. 0.075
B. 7.5
C. 750
D. 7500

M5M3a

24. $3,067 \times 3.4 =$ ______

A. 8,427.80
B. 10,427.80
C. 100,427.80
D. 9,427.80

M5N3c

25. Find the area.

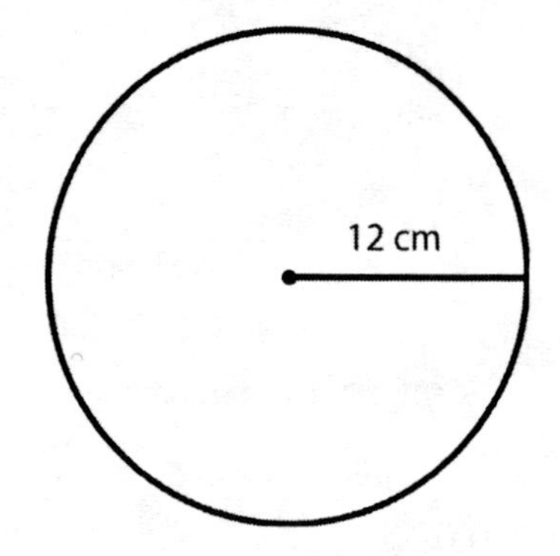

A. 452.16 cm^2
B. 144 cm^2
C. 4521 cm^2
D. 45.2 cm^2

M5M1e

26. $987 \div 8.6 =$ ______

A. 114.77
B. 110.7
C. 112
D. 116

M5N3d

27. Find the area of the irregular polygon illustrated below.

4 units

9 units

4 units

9 units

A. 56 sq. units
B. 36 sq. units
C. 20 sq. units
D. 81 sq. units

M5M1f

28. $1,528 \times 67 =$ ______

A. 112,376
B. 112,372
C. 102,367
D. 102,376

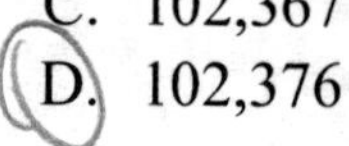

M5N3d

29. Find the quotient. $\frac{6,540}{12}$

A. 41
B. 545
C. 504
D. 501

M5N4a

30. Find the volume.

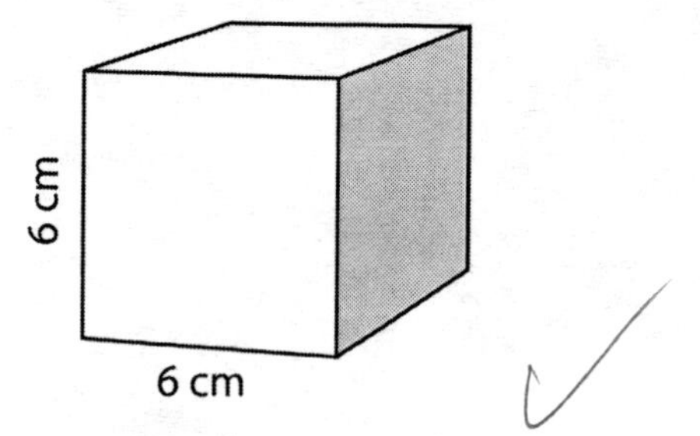

A. 36 cm
B. 36 cm^3
C. 216 cm^3
D. 216 cm

M5M4d

31. $816 \div 0.11 =$ ______

A. 89.76
B. 7,418.18
C. 8,418.18
D. 418

M5N3d

32. Find the volume of a box with the following measurements:

$l = 8$ feet
$w = 3.1$ feet
$h = 4$ feet

A. 98.2 ft^2
B. 24.8 ft
C. 99.2 ft^3
D. 96.4 ft

M5M4d

33. $1,821 \div 3 =$ _____

A. 67
B. 1,818
C. 607
D. 507

M5N4a

34. Find the volume of a cube, each edge has a length of 8 cm.

A. 64 cm
B. 64 cm^3
C. 512 cm^3
D. 512 cm

M5M4d

35. Simplify the following fraction.

$\frac{18}{24} =$

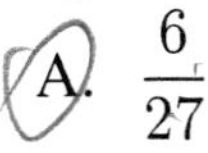

A. $\frac{1}{2}$

B. $\frac{1}{3}$

C. $\frac{3}{4}$

D. $\frac{1}{4}$

M5N4c

36. Find the quotient: $216 \div 2 =$

A. 104
B. 108
C. 118
D. 98

M5N4a

37. $\frac{1}{2} \times \frac{5}{5} =$

A. $\frac{1}{2}$

B. $\frac{7}{9}$

C. $\frac{30}{20}$

D. $\frac{20}{10}$

M5N4b

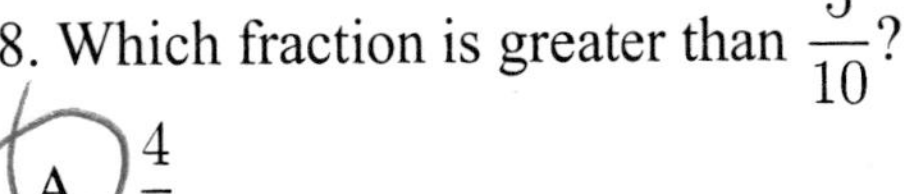

38. Which fraction is greater than $\frac{5}{10}$?

A. $\frac{4}{6}$

B. $\frac{2}{8}$

C. $\frac{2}{5}$

D. $\frac{2}{7}$

M5N4f

39. $\frac{3}{9}$ is equivalent to which fraction?

A. $\frac{6}{27}$

B. $\frac{1}{3}$

C. $\frac{6}{12}$

D. $\frac{1}{4}$

M5N4c

40. $1\frac{2}{3} - \frac{2}{5} =$

A. $1\frac{1}{3}$

B. $\frac{4}{15}$

C. $1\frac{4}{15}$

D. $1\frac{1}{15}$

M5N4g

41. $3\frac{3}{16} - \frac{5}{8} =$

A. $1\frac{3}{4}$

B. $2\frac{9}{16}$

C. $3\frac{9}{16}$

D. $3\frac{7}{16}$

M5N4g

42. $12\frac{1}{4} - 7\frac{1}{6} =$

A. $5\frac{1}{4}$

B. $5\frac{1}{12}$

C. $5\frac{1}{6}$

D. $5\frac{1}{2}$

M5N4g

43. Estimate: $267 \times 42 =$

A. 8,000
B. 12,000
C. 15,000
D. 10,000

M5N4i

44. Estimate: $16,144 \div 4,211 =$

A. 40
B. 400
C. 4
D. 0.4

M5N4h

45. $3\frac{1}{2}$ is equivalent to _____.

A. $3\frac{1}{3}$

B. 3.5

C. 0.33

D. 7.2

M5N4h

46. 1.6 is equivalent to _____.

A. $1\frac{3}{5}$

B. $1\frac{1}{3}$

C. $2\frac{1}{5}$

D. $1\frac{6}{100}$

M5N4h

47. $\frac{1}{8}$ is equivalent to _____.

A. 0.0125
B. 1.25
C. 0.125
D. 0.25

M5N4h

Use the following diagram to answer questions 48 and 49.

Transportation to School

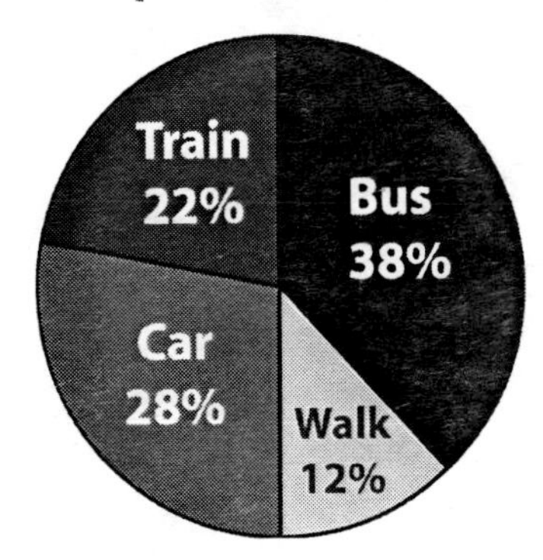

Total # of Students: 1200

48. How many students walk to school?

A. 336
B. 150
C. 144
D. 120

M5D1b

49. $\frac{3}{4}$ of all students who ride the bus, eat breakfast in school. How many students eat breakfast in school?

A. 456
B. 114
C. 264
D. 342

M5D1b

50. $\frac{75}{100} = $ _____.

A. 750
B. $\frac{25}{50}$
C. 7.5
D. $\frac{3}{4}$

M5N4h

51. Billie Jean purchased a sweater for $35.00. The next day, the sweater she purchased was on sale offering a 20% discount. How much money would she have saved, if she purchased the sweater the following day?

A. $28.00
B. $15.00
C. $7.00
D. $17.00

M5P2

52. Estimate: $81,006 \div 41 =$

A. 1,841.05
B. 2,000
C. 1,600
D. 1,700

M5N4i

53. $\frac{1}{4} \times \frac{2}{3} =$

A. 2
B. 12%
C. 25%
D. $\frac{1}{6}$

M5N4d

54. If $x = 2$, then $4x + 7 =$

A. 11
B. 28
C. 15
D. 8

M5A1b

55. Which amount is the biggest?

A. 2 cups
B. 8 fluid ounces
C. $\frac{1}{2}$ pint
D. $\frac{1}{4}$ quart

M5M3b

56. 146 pints = ______ cups

A. 36.50
B. 292
C. 18.25
D. 73

M5M3b

57. Estimate volume

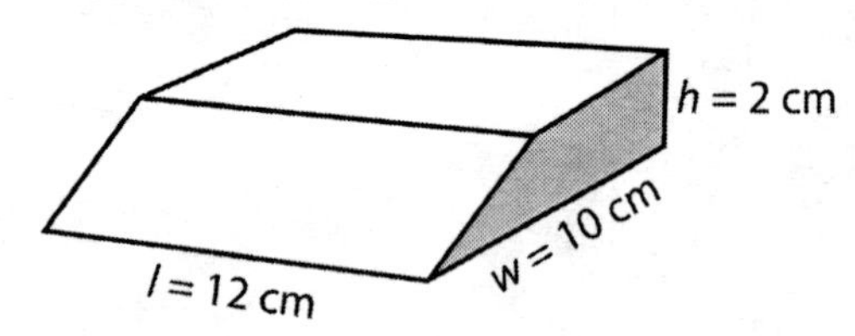

A. 20 cm^3
B. 240 cm^3
C. 265.61 cm^3
D. 19.7 cm^3

M5M4e

58. If the volume of a container is 210 cm^3, how many mLs will the container hold?

A. 105
B. 210
C. 21
D. 2.1

M5M4f

59. Figure $ABCD$ is a parallelogram. $\overline{AB}$ is congruent to ________.

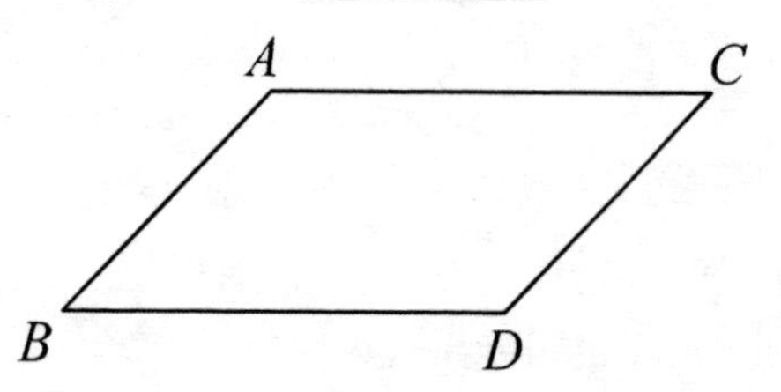

A. $\overline{CD}$
B. $\overline{BD}$
C. $\overline{AC}$
D. $\overline{AD}$

M5G1

60. Find the diameter of this circle.

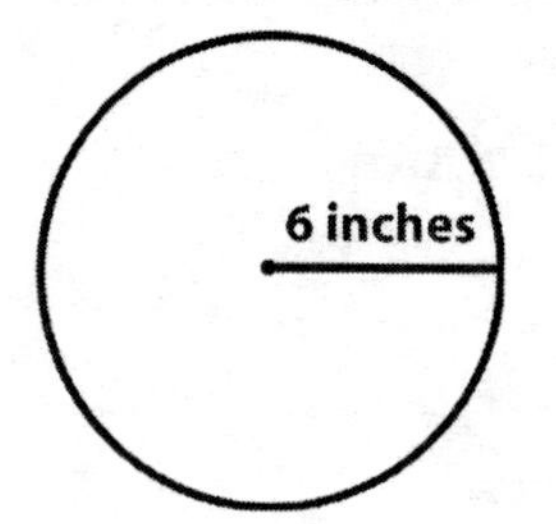

A. 12 inches
B. 18.84 inches
C. 36 inches
D. 37.68 inches

M5G2

61. Find the circumference of a circle with a radius of 16 cm.

A. 50.24 cm
B. 50 cm^2
C. 100.48 cm
D. 803.84 cm

M5G2

62. Translate "eighty-four less the product of six and seven" into an algebraic expression?

A. $(6 \times 7) - 84$
B. $(6 \times 7)(84)$
C. $84 - (6 \times 7)$
D. $84 \times (6 + 7)$

M5A1a

63. If $x = 3$, then $10x - 5 =$

A. 25
B. 5
C. 8
D. 30

M5A1b

64. Tiffany purchased a shirt for $15.35, plus 7% sales tax. Which expression would be used to determine the total Tiffany paid?

A. $15.35 + 0.07($15.35)
B. $15.35 + 0.07
C. $15.35 – 0.07($15.35)
D. $15.35

M5A1a

Use the graph below to answer question 65 and 66.

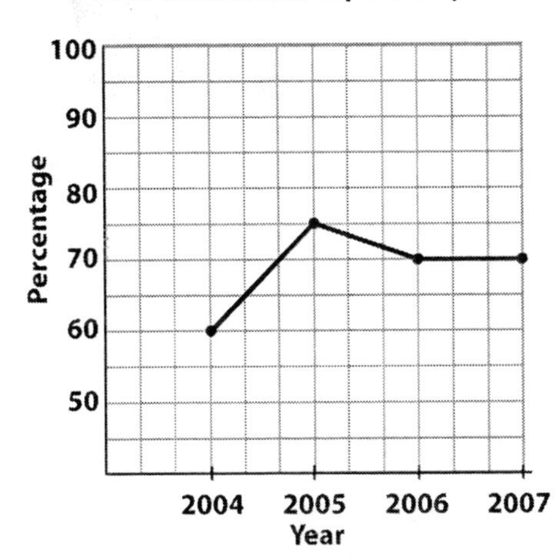

65. If 1,000 students are surveyed annually, how many students did **not** participate in field trips in 2004?

A. 600
B. 250
C. 400
D. 300

M5D1a

66. Which year showed the highest participation?

A. 2006
B. 2004
C. 2007
D. 2005

M5D1a

67. Jackie bought 4 books totaling $28.60. If each book cost the same amount, how much did one book cost?

A. $6.75
B. $7.15
C. $14.30
D. $7.00

M5P1a

68. Keith had $7.00 remaining, after he left the amusement park. At the park, he purchased 2 stuffed animals, 4 bottles of water, and one lunch special. How much money did Keith have at the beginning of the trip?

Item	Cost per unit
Stuffed Animal	$4.50
Bottled Water	$4.00
Lunch Special	$8.00

A. $30.00
B. $40.00
C. $32.00
D. $42.00

M5P1b

69. Madison had 48 dolls. She gave $\frac{1}{4}$ to her sister and donated $\frac{1}{3}$ to the local children's hospital. How many dolls did Madison keep?

A. 36
B. 32
C. 21
D. 20

M5P1a

70. How much change will Mr. Peterkin receive from $40.00, if he purchased 4 calculators and two notebooks on Wednesday?

Calculators	$6.75
Notebooks	$2.75
Pencils	$0.50

10% Discount Wednesdays only!

A. $10.00
B. $10.75
C. $3.25
D. $7.50

M5P1b

Practice Test 2

1. In this set of numbers, which numbers are the prime numbers?

   2 3 4 5 6 7

   A. 2, 3, 4, 5, 6, and 7 only
   B. 2, 3, 5, and 7 only
   C. 3, 5, and 7 only
   D. 3 and 7 only

   M5N1a

2. 2 and 7 are both factors of which of the following numbers?

   A. 9
   B. 21
   C. 27
   D. 28

   M5N1b

3. Which of the following number is not a multiple of 3?

   A. 6
   B. 9
   C. 13
   D. 15

   M5N1b

4. What is $5 \div 0$?

   A. 0
   B. 5
   C. 10
   D. You cannot divide by 0.

   M5N1c

5. In the number 356.7, what number is in the hundreds place?

   A. 3
   B. 5
   C. 6
   D. 7

   M5N2a

6. Zelie has had poison ivy 3 times. Louis boasts, "I've had poison ivy 10 times more than you!" If this is true, how many times has Louis had poison ivy?

   A. 0.3 times
   B. 3 times
   C. 30 times
   D. 300 times

   M5N2b

7. Which of the following equations is correct?

   A. $\frac{3}{2} = 3 \div 2$
   B. $\frac{3}{2} = 3 \times 2$
   C. $\frac{3}{2} = 2 \div 3$
   D. $\frac{3}{2} = 3 - 2$

   M5N4a

8. $\frac{3}{6}$ of Megan's friends are coming to her slumber party. She simplifies the fraction. Which of the following algebraic statements is true?

   A. $\frac{3}{6} = \frac{1}{2}$
   B. $\frac{3}{6} > \frac{1}{2}$
   C. $\frac{3}{6} < \frac{1}{2}$
   D. $3 \div 3 = \frac{1}{2}$

   M5N4b

9. What is $2.4 \div 0.6$?

A. 0.04
B. 0.4
C. 4
D. 40

M5N3c

10. There are 60 students in Haley's grade, and 40 of them watched the Super Bowl this year. Which of these fractions is equivalent to the fraction of the class that watched the Super Bowl?

A. $\frac{2}{6}$
B. $\frac{4}{60}$
C. $\frac{4}{3}$
D. $\frac{2}{3}$

M5N4c

11. Which of these inequalities is correct?

A. $\frac{1}{2} < \frac{2}{3}$
B. $\frac{3}{4} < \frac{2}{3}$
C. $\frac{3}{4} < \frac{1}{2}$
D. $\frac{3}{4} = \frac{1}{2}$

M5N4f

12. What is 1.1×8?

A. 0.8
B. 8
C. 8.8
D. 88

M5N3c

13. Which of these diagrams shows a carton of eggs that is $\frac{1}{3}$ full?

A. ●●●●○○ / ●●●●○○
B. ●●●●○○ / ○○○○○○
C. ●○○○○○ / ●●○○○○
D. ●○●○●○ / ○●○●○●

M5N4e

14. Mia and Antonio host a pizza party. As they are cleaning up, Mia collects $1\frac{5}{8}$ leftover pizzas and Antonio collects $2\frac{1}{2}$ leftover pizzas. How many total leftover pizzas do they have?

A. $3\frac{1}{8}$ pizzas
B. $4\frac{1}{8}$ pizzas
C. $3\frac{6}{10}$ pizzas
D. $1\frac{1}{8}$ pizzas

M5N4g

15. What is the best estimate of $\frac{1}{5}$ of 497?

A. 800
B. 500
C. 50
D. 100

M5N4i

16. Miss Madden asks her class which season they prefer: Spring, Summer, Fall, or Winter. The circle graph below shows their results.

Miss Madden's Class, Favorite Seasons

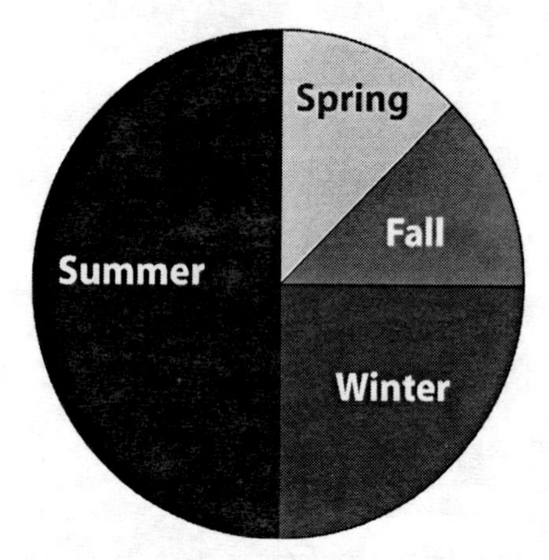

Which of the following tables shows a correct interpretation of this circle graph?

A.

Season	Percentage
Spring	25%
Summer	25%
Fall	25%
Winter	25%

B.

Season	Percentage
Spring	25%
Summer	50%
Fall	25%
Winter	25%

C.

Season	Percentage
Spring	25%
Summer	50%
Fall	12.5%
Winter	12.5%

D.

Season	Percentage
Spring	12.5%
Summer	50%
Fall	12.5%
Winter	25%

M5N5b

17. Suzanne's last exam score is represented by the 10×10 grid below.

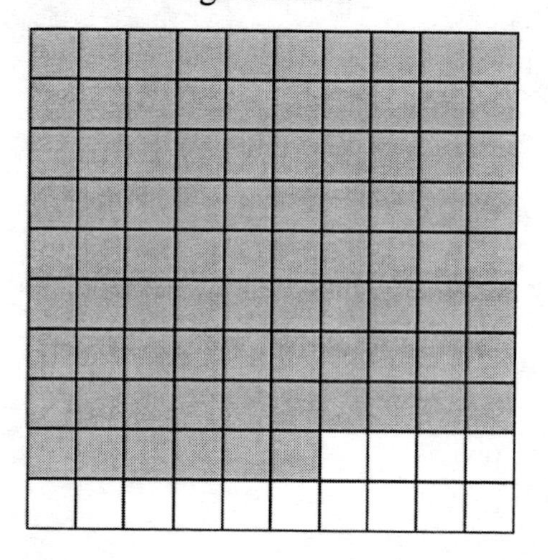

What did Suzanne score on the exam?

A. 68%
B. 76%
C. 86%
D. 96%

M5N5a

18. Adam, Denny, Kim, and Ali split the cost of a \$30 boat rental. The cost to each of them can be represented by $\$\frac{30}{4}$. How much will each of them pay, in decimal form?

A. \$7.50
B. \$8.00
C. \$8.50
D. \$9.00

M5N4h

19. Tayshaun has a 2 L bottle of soda. How many mL (milliliters) of soda does he have?

A. 0.2
B. 20
C. 200
D. 2000

M5M3b

20. Setrak, Caylene, and Marley stand in a triangle with a base b of 10 ft and a height h of 6 ft. The formula for the area of a triangle is $A = \frac{1}{2} \times b \times h$.

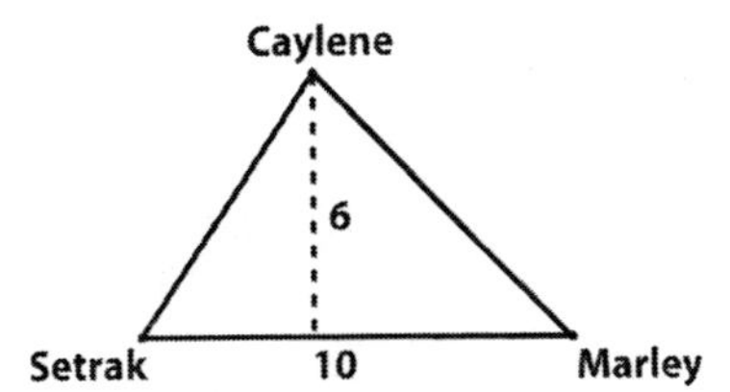

What is the area of the triangle?

A. 5 ft^2
B. 30 ft^2
C. 60 ft^2
D. 120 ft^2

M5N4d

21. Mike's glow-in-the-dark frisbee is a circle with radius 5. The formula for the area of a circle is $A = \pi \times r^2$. What is the area of Mike's glow in the dark frisbee? Use $\pi = 3.14$.

A. 5
B. 15.7
C. 25
D. 78.5

M5M1e

22. Lucas has a Rubik's cube with each edge 3 inches. What is the volume of the cube? Use the formula $V = e^3$.

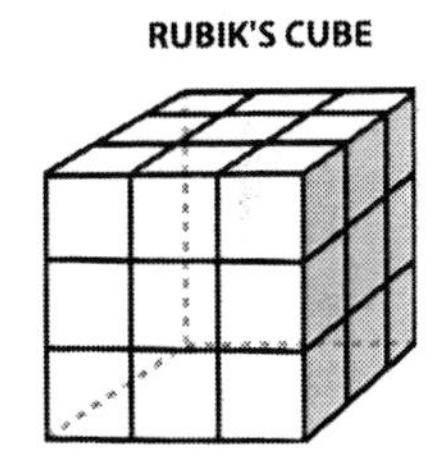

A. 27 in^3
B. 24 in^3
C. 9 in^3
D. 6 in^3

M5M4d

23. Find the area of the pentagon below by dividing it into a triangle and a square. Use the formula for the area of a triangle, $A = \frac{1}{2} \times b \times h$, and the formula for the area of a square, $A = s^2$.

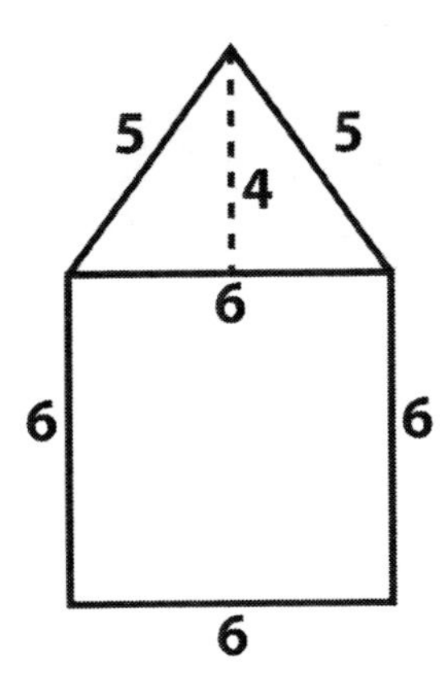

A. 46
B. 48
C. 51
D. 60

M5M1f

24. Amanda buys a 3 gallon tub of vanilla ice cream at the BuyInBulk store. How many quarts of ice cream does she have?

A. 0.75 quarts
B. 6 quarts
C. 12 quarts
D. 24 quarts

M5M3b

25. J.O. has an aquarium for his pet piranha, Paco. Which of these units could be used to measure the volume of the aquarium?

A. cm
B. cm^2
C. cm^3
D. cm^4

M5M4b

26. Lisha has a box of tissues that has a height of 6 inches, a length of 10 inches, and a width of 6 inches. What is the volume of the box?

A. 36 in^3
B. 22 in^3
C. 60 in^3
D. 360 in^3

M5M4e

27. Look at the triangle below.

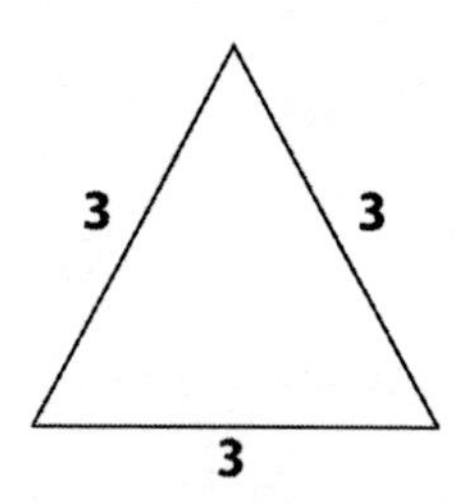

Which of these triangles is congruent to the triangle above?

A.

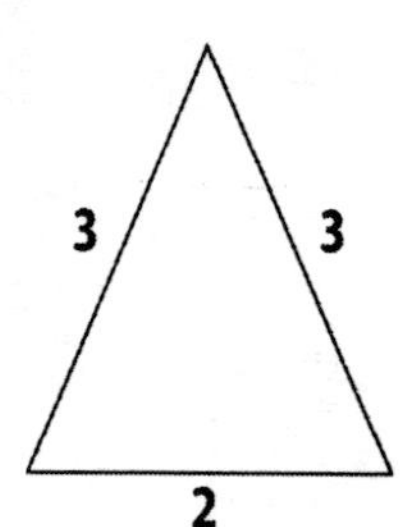

B.

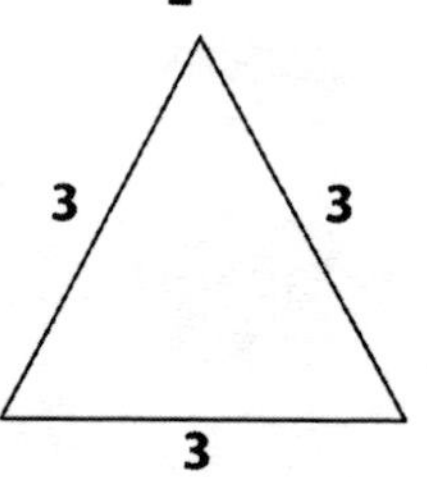

C.

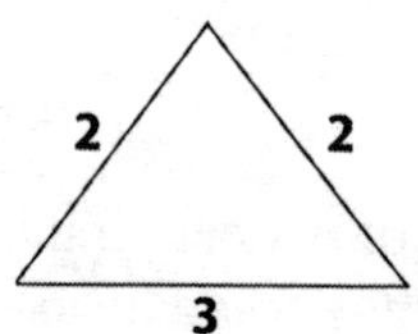

D.

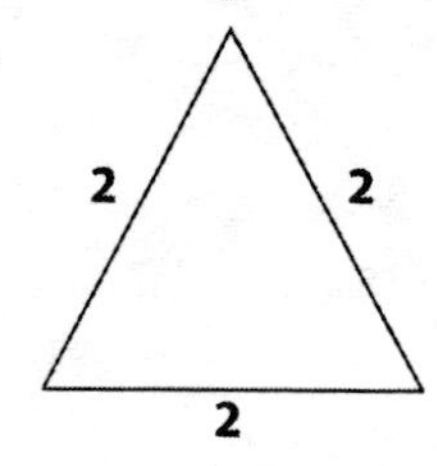

M5G1

28. Which of these relationships in a circle is correct?

A. Diameter $\times \pi =$ Circumference
B. Circumference $\times \pi =$ Diameter
C. Circumference $\times$ Diameter $= \pi$
D. Diameter $\div$ Circumference $= \pi$

M5G2

29. Thelma and Louise go to a ball game. Sodas cost $1.50. Which algebraic expression represents this situation, if n is the number of sodas that they bought?

A. $1.5n$
B. $\frac{n}{1.5}$
C. $n - 1.5$
D. $n + 1.5$

M5A1a

30. Cpt. Tollefson has some money in the stock market. The value of his money doubled, and then he lost $10. Which of these expressions represents the amount of money he has now, if x is the amount of money he had originally?

A. $x + 2 - 10$
B. $2x + 10$
C. $2x - 10$
D. $\frac{2x}{10}$

M5A1a

31. What is $3x - 9$, when $x = 4$?

A. 2
B. 3
C. 4
D. 5

M5A1b

32. The size of Mr. McKinley's stamp collection is given by the equation $x + 7$, where x is the number of stamps he had last week. If he had 74 stamps last week, what is the size of his stamp collection now?

A. 7 stamps
B. 67 stamps
C. 74 stamps
D. 81 stamps

M5A1b

33. Amedee does a long division problem and ends up with the decimal 0.6. Which of the following is a fractional equivalent to 0.6?

A. $\frac{0}{6}$
B. $\frac{1}{6}$
C. $\frac{3}{5}$
D. $\frac{2}{3}$

M5N4h

34. The bar graph below shows the number of cookies eaten by each member of the Hanning family.

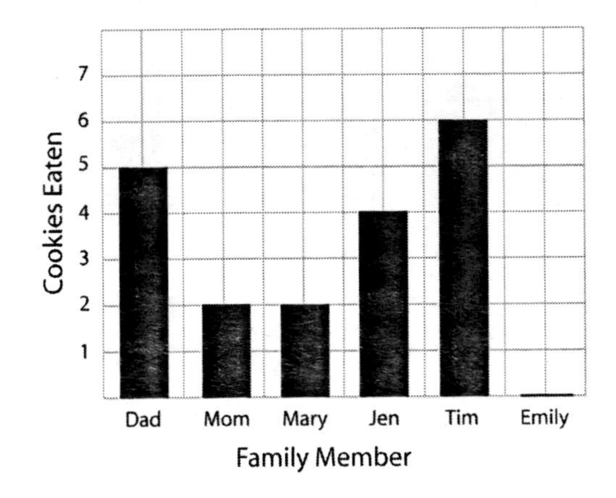

How many cookies did Jen eat?

A. 2
B. 4
C. 5
D. 6

M5D1a

35. The chart below shows the amount of rainforest (by acre) in the small tropical country of Bixenta.

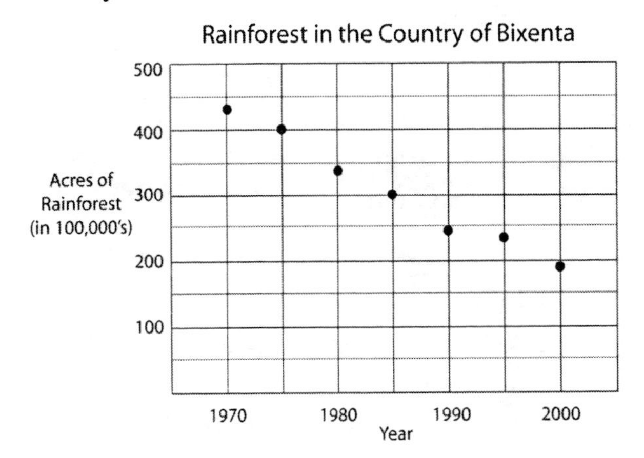

Which of these statements appears to be true about the amount of rainforest, based on the graph?

A. The amount of rainforest is decreasing.
B. The amount of rainforest is increasing.
C. The amount of rainforest is remaining constant.
D. This question cannot be answered using this graph.

M5D1a

36. If you wanted to show what percentage of your time you spent on different activities during the day, which kind of graph would be the best to display it?

A. bar graph
B. line graph
C. pictograph
D. circle graph

M5D1b

37. To show the change in the size of Atlanta's population over time, the best kind of graph would be

A. a circle graph.
B. a line graph.
C. a histogram.
D. a number line.

M5D1b

38. Martin wants to show his company's yearly profits compared to their four biggest competitors' profits. Which kind of graph should he choose?

A. a bar graph
B. a circle graph
C. a pictograph
D. a line graph

M5D1b

39. Dwayne measures the inside of his basketball hoop as 18 inches across. If this is the diameter, which of the following is the best estimate of the inside circumference of the hoop?

A. 6 inches
B. 27 inches
C. 54 inches
D. 81 inches

M5G2

40. Patrice's mom works as a tax consultant. The line graph below shows the number of cups of coffee her mom consumes by month. (April is tax month.)

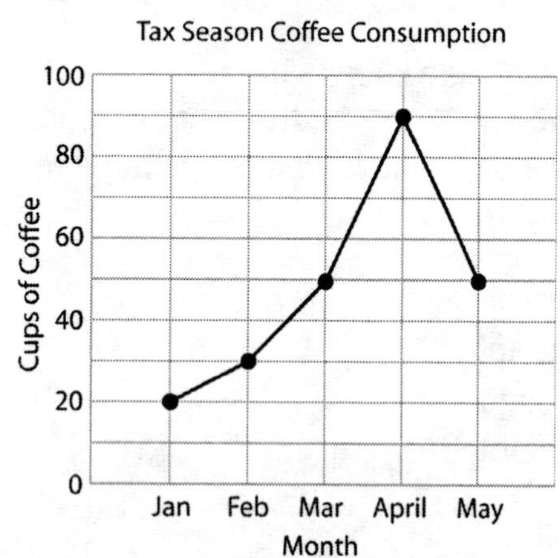

How many more cups did Patrice's mom consume in April than in February?

A. 30
B. 60
C. 70
D. 90

M5D1a

41. Trapezoid $ABCD$ is congruent to Trapezoid $WXYZ$.

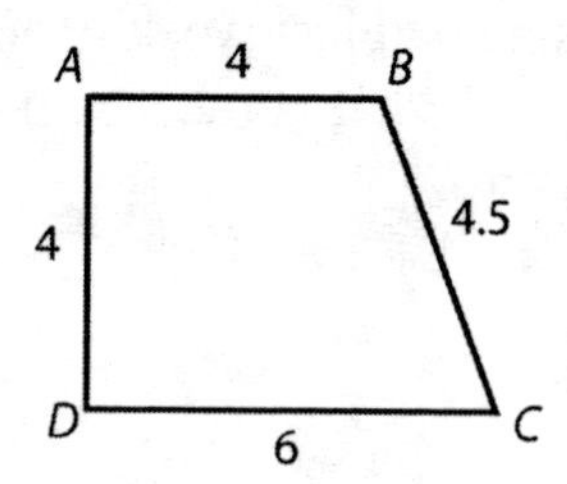

Which of the following is the correct labeling of Trapezoid $WXYZ$?

A.
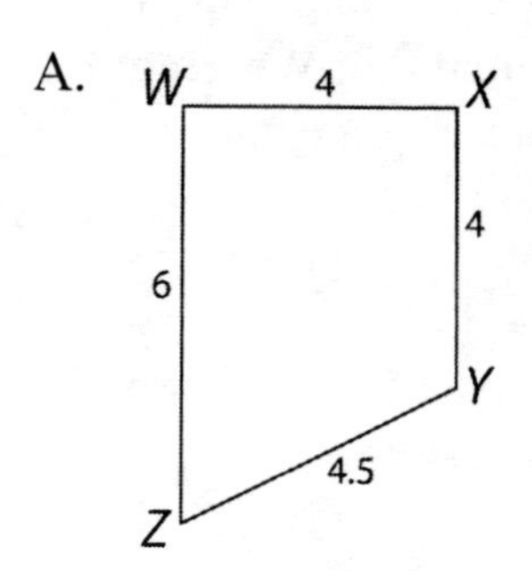

B.
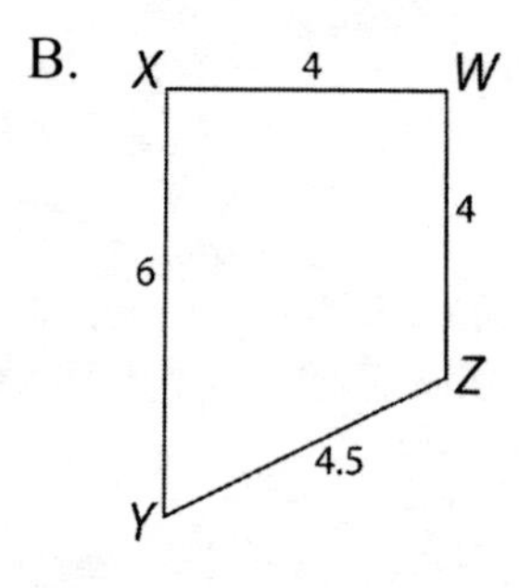

C.
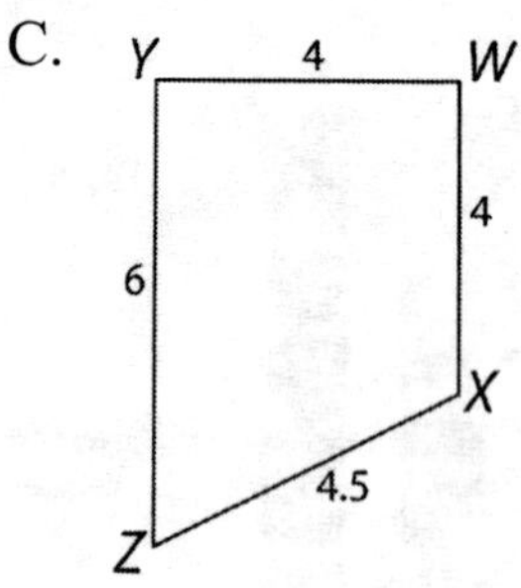

D.
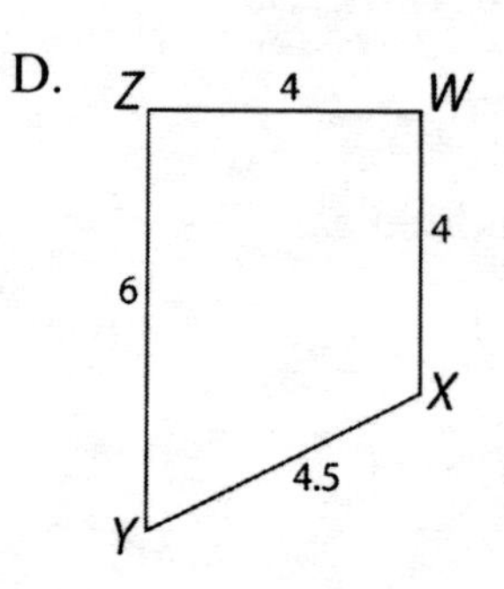

M5G1

42. The graph below shows the average monthly high temperature in Atlanta for the first six months of the year.

Jan	Feb	Mar	April	May	June
53°	58°	74°	78°	84°	87°

Which of the following line graphs accurately displays this information?

A. 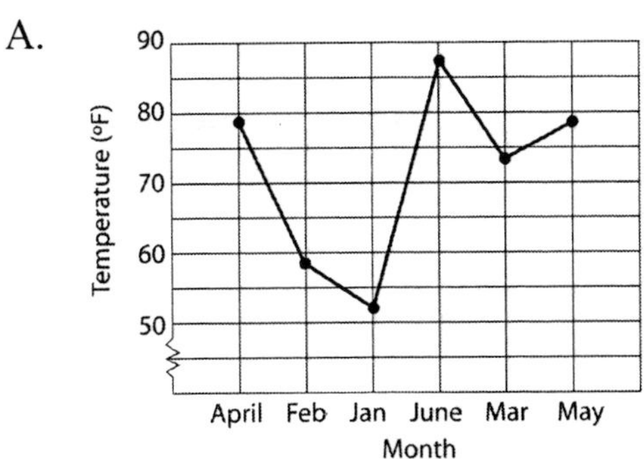

B. 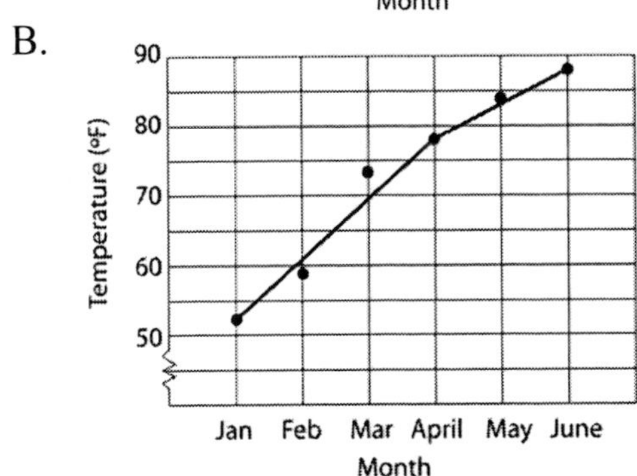

C. 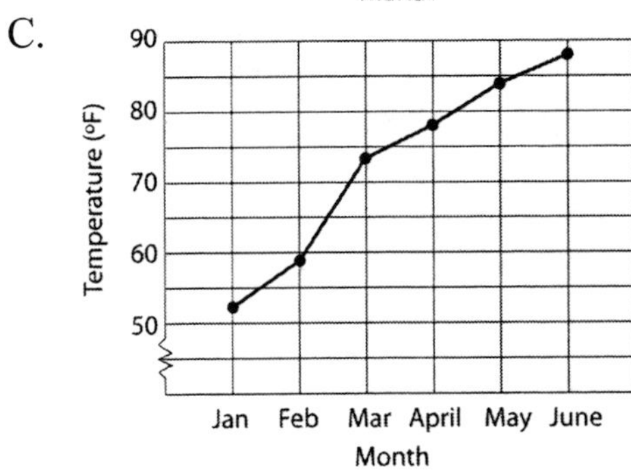

D. 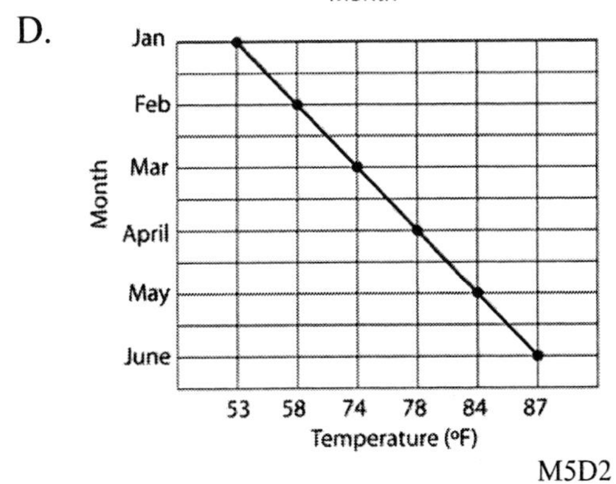

M5D2

43. For a class project, Feenstra recorded the last $100 she spent.

Feenstra's Spending

Category	Amount
Movies	$40
Food	$30
Clothing	$20
Library Fine	$10

Which of the following circle graphs accurately depicts Feenstra's spending?

A. 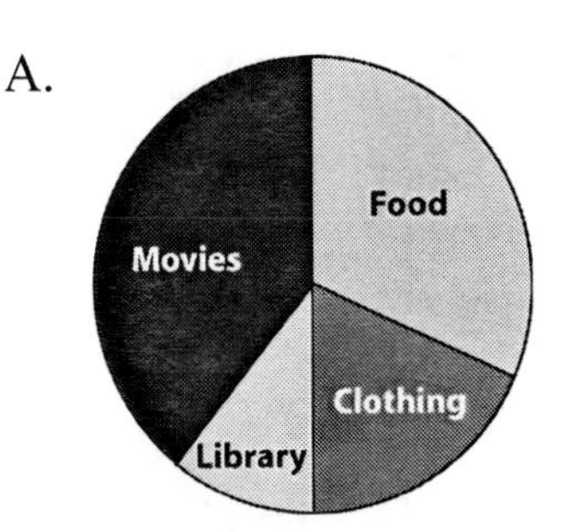

B. 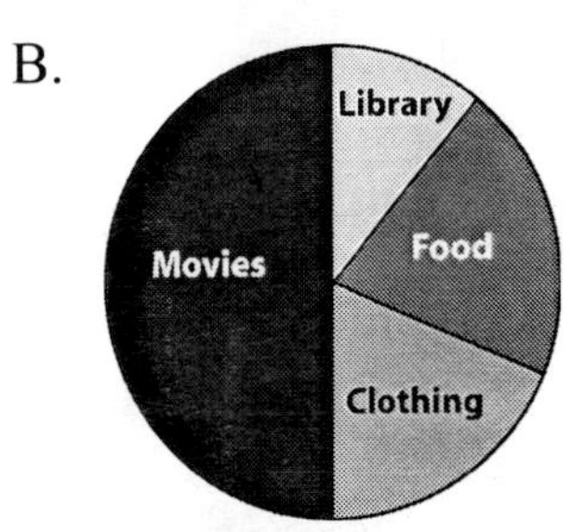

C. 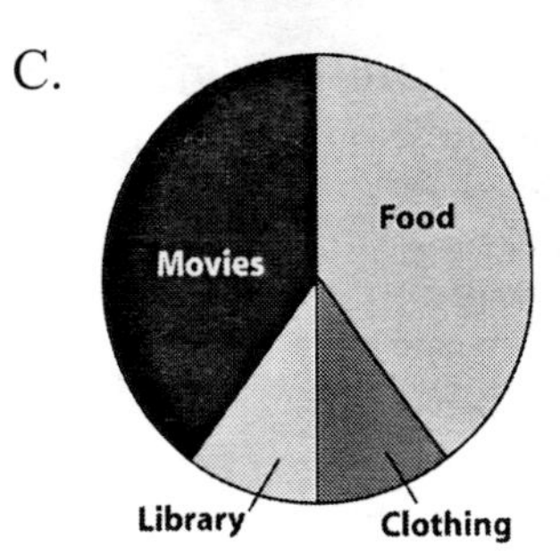

D. 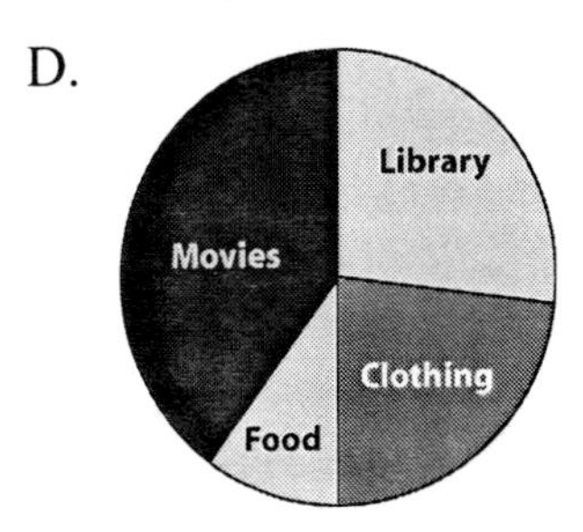

M5D2

44. The 5th grade band currently has 32 members, and Mr. Woodward announces that there will be new students joining next week. Which of the following expressions represents this situation?

A. $32 \times x$
B. $32 \div x$
C. $32 - x$
D. $32 + x$

M5A1a

45. Which of the following is the best estimate of the area of the rectangle below?

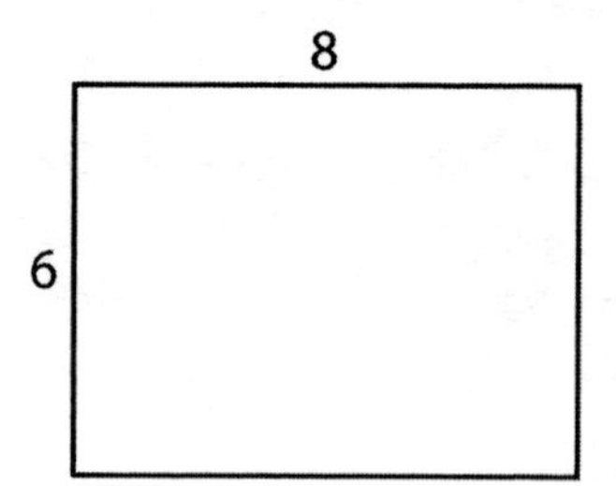

A. 40
B. 50
C. 60
D. 70

M5M1a

46. The area of rectangle $ABCD$ is given by $A = b \times h$. What is the formula for the area of the triangle ADC?

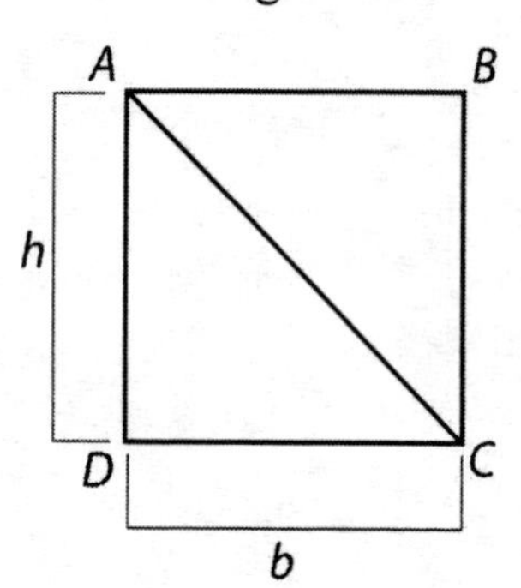

A. $\frac{1}{2} \times b \times h$
B. $b \times h$
C. $2 \times b \times h$
D. $b^2 \times h$

M5M1c

47. Rectangle $MLNO$ has base 6 and height 4. Parallelogram $PQRS$ has base 6 and height 4. What is the relationship of their areas?

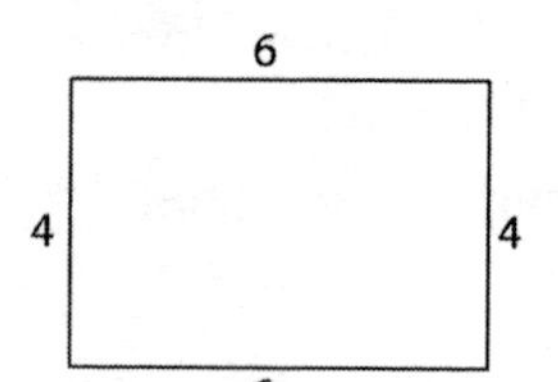

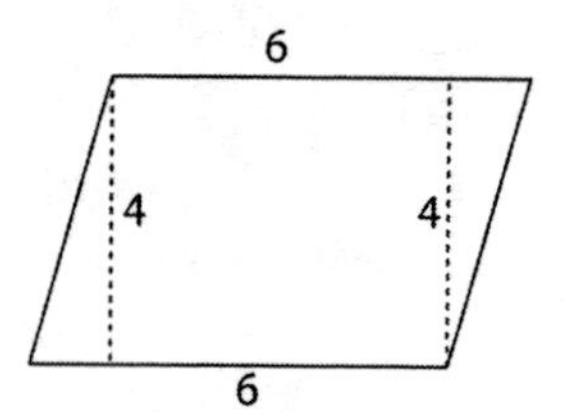

A. Area $MLNO >$ Area $PQRS$
B. Area $MLNO <$ Area $PQRS$
C. Area $MLNO =$ Area $PQRS$
D. Area $MLNO =$ Area $PQRS \times 2$

M5M1b

48. What is the area of the "L" shape below?

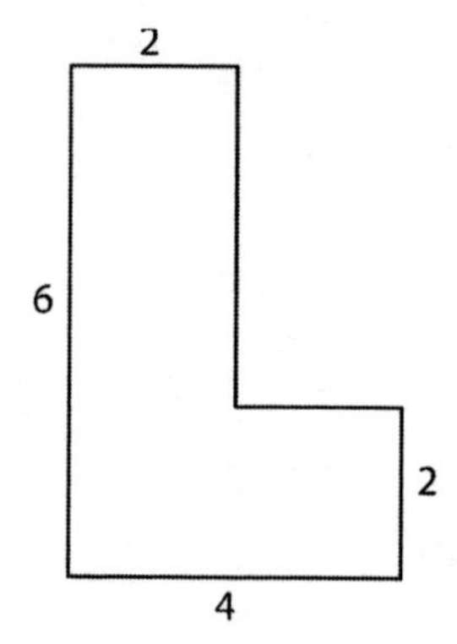

A. 28
B. 24
C. 20
D. 16

M5M1f

49. Amber has 3 full pint-size containers of orange juice. Ashja has 2 full quart-size containers of orange juice. Which girl has more orange juice?

A. Amber
B. Ashja
C. They have the same amount
D. The amounts cannot be compared without more information.

M5M3b

50. Most American cars have gasoline tanks that can hold 10 to 30

A. gallons.
B. pints.
C. quarts.
D. fluid ounces.

M5M3a

51. Yusef has a box of crackers with dimensions $9 \times 4 \times 4$ (all inches). What is the volume of this box?

A. 17 in^3
B. 23 in^3
C. 72 in^3
D. 144 in^3

M5M4d

52. The formula for the area of a parallelogram is $A = b \times h$. What is the area of the parallelogram below?

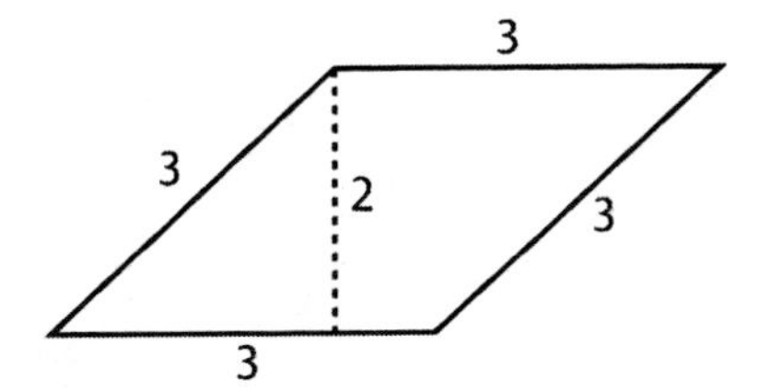

A. 12
B. 9
C. 6
D. 3

M5M1d

53. What is $23 \div 1$?

A. 23
B. 24
C. 22
D. You cannot divide by 1.

M5N1c

54. In the number 24.68, which number is in the tenths place?

A. 2
B. 4
C. 6
D. 8

M5N2a

55. Which of the following fractions is equal to $\frac{12}{36}$ completely simplified?

A. $\frac{6}{12}$
B. $\frac{4}{12}$
C. $\frac{3}{9}$
D. $\frac{1}{3}$

M5N4c

56. Tip on a meal at a fancy restaurant can be as high as 20% or 0.2 times the cost of the meal. John Paul's food costs \$6.50. How much should his tip be?

A. \$0.13
B. \$1.30
C. \$1.40
D. \$12.00

M5N3c

57. Which of the following sets of numbers has no even number?

A. $2, 3, 5, 7, 11, 13, 17$
B. $3, 6, 9, 12, 15, 18, 21$
C. $4, 6, 8, 9, 10, 12, 14$
D. $3, 9, 15, 21, 27, 33$

M5N1a

58. What is $\frac{4}{5} - \frac{1}{3} = ?$

A. $\frac{3}{2}$

B. $\frac{3}{5}$

C. $\frac{7}{15}$

D. $\frac{12}{15}$

M5N4g

59. Therese has a tin of cookies with volume 2400 cm^3. She has a large number of square cookies that she would like to put in the tin. If she does not crush the cookies, what volume of cookies will she be able to fit into tin?

A. Volume of cookies = 2400 cm^3
B. Volume of cookies < 2400 cm^3
C. Volume of cookies > 2400 cm^3
D. Volume of cookies = 2400 $cm^3 \div 4$

M5M4f

60. A backyard is 160 feet wide. What is the width of the field in yards?

A. 50 yards
B. 53 yards
C. $53\frac{1}{3}$ yards
D. $53\frac{2}{3}$ yards

M5M3b

61. Joshua, quarterback for the Patriots, held his hands at an angle to receive the "snap" from the center.

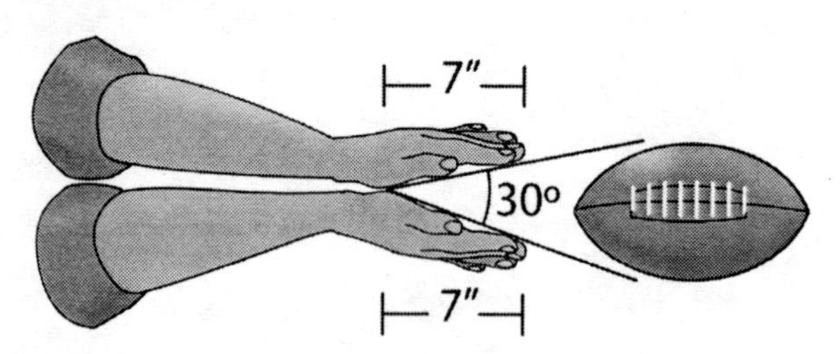

Which of the angles below is congruent to the angle made by Joshua's hands?

A.
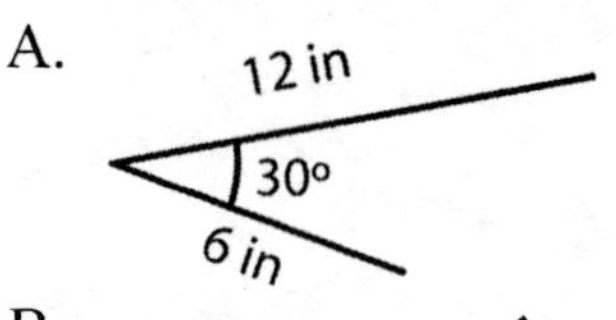

B.
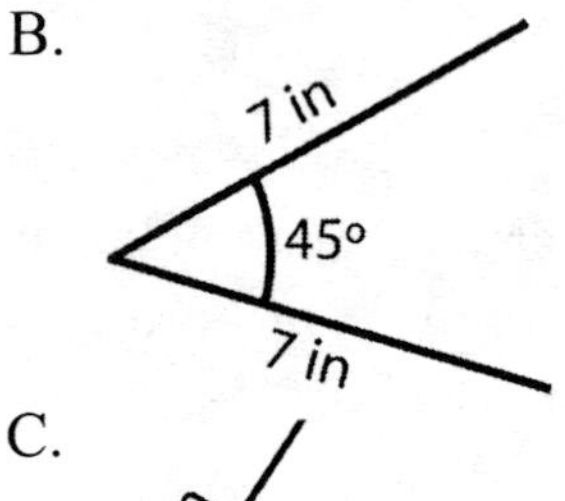

C.
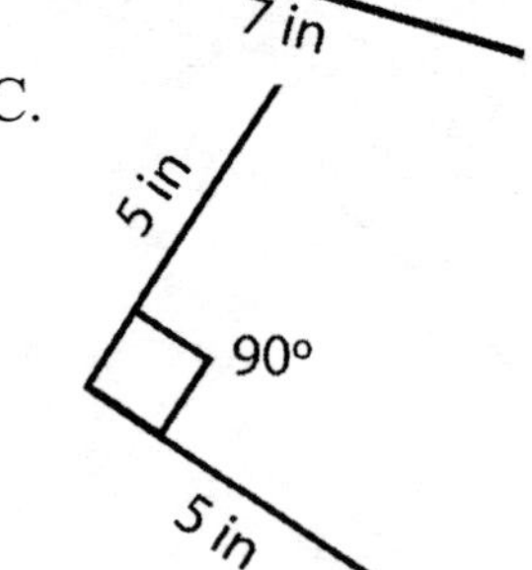

D.
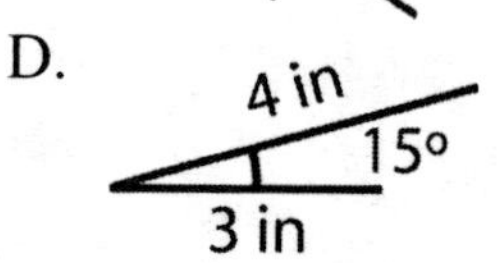

M5G1

62. Which of the following can be used to compute $\frac{4}{5} + \frac{5}{6}$?

A. $\frac{4+5}{5+6}$

B. $\frac{4 \times 6}{5 \times 6} + \frac{5 \times 5}{6 \times 5}$

C. $\frac{4}{5 \times 1} + \frac{5}{6 \times 2}$

D. $\frac{4 \times 5}{5 \times 5} + \frac{5 \times 6}{6 \times 6}$

M5N4g

63. Find the area of the shaded area below. (Diagram not drawn to scale.)

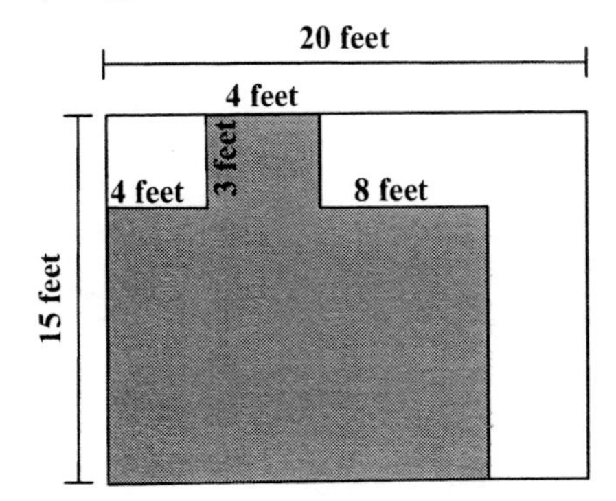

A. 204 ft^2
B. 264 ft^2
C. 276 ft^2
D. 300 ft^2

M5M1f

64. What is the area of a circle with a radius of 7 cm? (Round to the nearest whole number)

A. 154 square cm
B. 196 square cm
C. 347 square cm
D. 616 square cm

M5M1e

65. The problem below was done on the calculator, but the decimal point was missing. Estimate to find the correct answer.

$795.6 \div 4.7$

A. 1.6927659
B. 16.927659
C. 169.27659
D. 1692.7659

M5N4i

66. Which of the following is a multiple of 6?

A. 14
B. 20
C. 24
D. 28

M5N1b

67. Which fraction is between $\frac{1}{10}$ and $\frac{1}{1000}$?

A. $\frac{2}{10}$

B. $\frac{3}{1000}$

C. $\frac{11}{100}$

D. $\frac{101}{1000}$

M5N4f

68. $4\frac{5}{6}$ written as a decimal is

A. 0.456
B. 4.56
C. $4.8\overline{3}$
D. $0.48\overline{3}$

M5N4h

69. 0.45 written as a fraction is

A. $\frac{9}{20}$

B. $\frac{5}{9}$

C. $\frac{4}{5}$

D. $\frac{1}{45}$

M5N4h

70. Which of the following is a factor of 26?

A. 13
B. 3
C. 6
D. 5

M5N1b

Index

ANSWER KEY

FOR

Mastering the Georgia 5th Grade CRCT in Mathematics

Erica Day

Colleen Pintozzi

Tanya Kelley

AMERICAN BOOK COMPANY

P. O. BOX 2638

WOODSTOCK, GA 30188-1383

TOLL FREE 1 (888) 264-5877 PHONE (770) 928-2834 FAX (770) 928-7483

Web site: www.americanbookcompany.com

The standards listed at the beginning of each chapter correspond to the following Georgia Performance Standards descriptions.

Numbers and Operations

M5N1. Students will further develop their understanding of whole numbers.

(A) Classify the set of counting numbers into subsets with distinguishing characteristics (odd/even, prime/composite).
(B) Find multiples and factors.
(C) Analyze and use divisibility rules.

M5N2. Students will further develop their understanding of decimal fractions as part of the base-ten number system.

(A) Understand place value
(B) Analyze the effect on the product when a number is multiplied by 10, 100, 1000, 0.1, and 0.01.

M5N3. Students will further develop their understanding of the meaning of multiplication and division with decimal fractions and use them.

(A) Model multiplication and division of decimal fractions by another decimal fraction.
(B) Explain the process of multiplication and division, including situations in which the multiplier and divisor are both whole numbers and decimal fractions.
(C) Multiply and divide with decimal fractions including decimal fractions less than one and greater than one.
(D) Understand the relationships and rules for multiplication and division of whole numbers also apply to decimal fractions.

M5N4. Students will continue to develop their understanding of the meaning of common fractions and compute with them.

(A) Understand division of whole numbers can be represented as a fraction ($a/b = a \div b$).
(B) Understand the value of a fraction is not changed when both its numerator and denominator are multiplied or divided by the same number because it is the same as multiplying or dividing by one.
(C) Find equivalent fractions and simplify fractions.
(D) Model the multiplication and division of common fractions.
(E) Explore finding common denominators using concrete, pictorial, and computational models.
(F) Use $<$, $>$, or $=$ to compare fractions and justify the comparison.
(G) Add and subtract common fractions and mixed numbers with unlike denominators.
(H) Use fractions (proper and improper) and decimal fractions interchangeably.
(I) Estimate products and quotients.

M5N5. Students will understand the meaning of percentage.

(A) Model percent on 10 by 10 grids.
(B) Apply percentage to circle graphs.

Measurement

M5M1. Students will extend their understanding of area of fundamental geometric plane figures.

(A) Estimate the area of fundamental geometric plane figures.
(B) Derive the formula for the area of a parallelogram (e.g., cut the parallelogram apart and rearrange it into a rectangle of the same area).
(C) Derive the formula for the area of a triangle (e.g. demonstrate and explain its relationship to the area of a rectangle with the same base and height).
(D) Find the areas of triangles and parallelograms using formulae.
(E) Estimate the area of a circle through partitioning and tiling and then with formula (let pi = 3.14). (Discuss square units as they apply to circles.)
(F) Find the area of a polygon (regular and irregular) by dividing it into squares, rectangles, and/or triangles and find the sum of the areas of those shapes.

M5M3. Students will measure capacity with appropriately chosen units and tools.

(A) Use milliliters, liters, fluid ounces, cups, pints, quarts, and gallons to measure capacity.
(B) Compare one unit to another within a single system of measurement (e.g., 1 quart = 2 pints).

M5M4. Students will understand and compute the volume of a simple geometric solid.

1.(A) Understand a cubic unit (u^3) is represented by a cube in which each edge has the length of 1 unit.
(B) Identify the units used in computing volume as cubic centimeters (cm^3), cubic meters (m^3), cubic inches (in^3), cubic feet (ft^3), and cubic yards (yd^3).
(C) Derive the formula for finding the volume of a cube and a rectangular prism using manipulatives.
(D) Compute the volume of a cube and a rectangular prism using formulae.
(E) Estimate the volume of a simple geometric solid.
(F) Understand the similarities and differences between volume and capacity.

Geometry

M5G1. Students will understand congruence of geometric figures and the correspondence of their vertices, sides, and angles.

M5G2. Students will understand the relationship of the circumference of a circle to its diameter is pi ($\pi \approx 3.14$).

Algebra

M5A1. Students will represent and interpret the relationships between quantities algebraically.

(A) Use variables, such as n or x, for unknown quantities in algebraic expressions.
(B) Investigate simple algebraic expressions by substituting numbers for the unknown.
(C) Determine that a formula will be reliable regardless of the type of number (whole numbers or decimal fractions) substituted for the variable.

Data Analysis

M5D1. Students will analyze graphs.

(A) Analyze data presented in a graph.
(B) Compare and contrast multiple graphic representations (circle graphs, line graphs, bar graphs, etc.) for a single set of data and discuss the advantages/disadvantages of each.

M5D2. Students will collect, organize, and display data using the most appropriate graph.

Process Standards

M5P1. Students will solve problems (using appropriate technology).

(A) Build new mathematical knowledge through problem solving.
(B) Solve problems that arise in mathematics and in other contexts.
(C) Apply and adapt a variety of appropriate strategies to solve problems.
(D) Monitor and reflect on the process of mathematical problem solving.

M5P2. Students will reason and evaluate mathematical arguments.

(A) Recognize reasoning and proof as fundamental aspects of mathematics.
(B) Make and investigate mathematical conjectures.
(C) Develop and evaluate mathematical arguments and proofs.
(D) Select and use various types of reasoning and methods of proof.

M5P3. Students will communicate mathematically.

(A) Organize and consolidate their mathematical thinking through communication.
(B) Communicate their mathematical thinking coherently and clearly to peers, teachers, and others.
(C) Analyze and evaluate the mathematical thinking and strategies of others.
(D) Use the language of mathematics to express mathematical ideas precisely.

M5P4. Students will make connections among mathematical ideas and to other disciplines.

(A) Recognize and use connections among mathematical ideas.
(B) Understand how mathematical ideas interconnect and build on one another to produce a coherent whole.
(C) Recognize and apply mathematics in contexts outside of mathematics.

M5P5. Students will represent mathematics in multiple ways.

(A) Create and use representations to organize, record, and communicate mathematical ideas.
(B) Select, apply, and translate among mathematical representations to solve problems.
(C) Use representations to model and interpret physical, social, and mathematical phenomena.

Diagnostic Test

Pages 1–10

1. A	8. C	15. D	22. A	29. A	36. B	43. D	50. B	57. B	64. B
2. A	9. D	16. B	23. A	30. C	37. C	44. A	51. A	58. C	65. B
3. A	10. B	17. C	24. A	31. B	38. A	45. C	52. C	59. C	66. D
4. B	11. C	18. C	25. B	32. B	39. B	46. A	53. B	60. C	67. D
5. A	12. D	19. A	26. A	33. B	40. B	47. A	54. C	61. B	68. B
6. B	13. B	20. C	27. B	34. D	41. A	48. B	55. B	62. B	69. C
7. C	14. D	21. B	28. B	35. A	42. A	49. C	56. A	63. A	70. A

Chapter 1 Whole Numbers and Number Sense

Page 14 Place Value: Greater Than One

1. Four thousand (4, 000)
2. Four thousand (4, 000)
3. Four ones (4)
4. Forty million (40, 000, 000)
5. Forty (40)
6. Four million (4, 000, 000)
7. Four hundred million (400, 000, 000)
8. Four thousand (4, 000)
9. 0
10. 7
11. 6
12. 8
13. 1
14. 2
15. 5
16. 4
17. five ones
18. three million
19. sixty
20. twenty thousand
21. six hundred million
22. one hundred thousand
23. eighty
24. twenty

Page 15 Even and Odd Numbers

1. e	3. o	5. o	7. o	9. o	11. o
2. e	4. o	6. o	8. e	10. e	12. o

Page 16 Practicing Adding

1. 10,791	4. 3,511	7. 12,745	10. 7,600	13. 3,413
2. 11,881	5. 4,849	8. 10,183	11. 9,436	14. 11,549
3. 17,507	6. 3,503	9. 10,909	12. 2,370	15. 5,736

Page 17 Adding Whole Numbers

1. 199
2. 7,469
3. 5,347
4. 443
5. 4,297
6. 2,951
7. 113
8. 635
9. 835
10. 167
11. 562
12. 175
13. 323
14. 943
15. 566
16. 360
17. 378
18. 268

Page 18 Practicing Subtraction

1. 5,016
2. 6,188
3. 1,257
4. 8,455
5. 1,668
6. 2,112
7. 3,244
8. 7,516
9. 2,863
10. 6,863
11. 14,992
12. 186
13. 45,309
14. 135
15. 6,505

Page 19 Subtracting Whole Numbers

1. 506
2. 5,728
3. 407
4. 431
5. 124
6. 564
7. 593
8. 527
9. 355
10. 435
11. 709
12. 1,019
13. 608
14. 675
15. 218
16. 2,093
17. 726
18. 369

Page 20 Multiplying Whole Numbers

1. 18,576
2. 32,648
3. 25,120
4. 24,444
5. 22,707
6. 6,156
7. 21,183
8. 25,424
9. 14,820
10. 71,910
11. 14,525
12. 46,312
13. 22,608
14. 20,748
15. 20,124
16. 21,432
17. 48,295
18. 20,026
19. 20,436
20. 43,648

Page 21 Divisibility Rules

1. 3
2. 5
3. 2, 5, 10
4. 2
5. 2, 3
6. 2, 3
7. 5
8. 2
9. 2, 5, 10
10. 2, 5, 10
11. 2, 3
12. 3

Page 22 Dividing Whole Numbers

1. 482 r8
2. 55 r19
3. 531 r3
4. 285 r2
5. 297 r23
6. 151 r35
7. 152 r22
8. 124 r22
9. 173 r20
10. 359 r30
11. 137 r14
12. 113
13. 307 r30
14. 121 r55
15. 143 r30
16. 2,142 r5

Page 23 Multiplying and Dividing by Multiples of Ten

1. 2,700	6. 126,100,000	11. 1,000	16. 3,500	21. 36,000
2. 356,000	7. 3,900	12. 20,000	17. 30	22. 40
3. 47,100,000	8. 42,000	13. 40	18. 5,000	23. 1,500
4. 371,400	9. 41,700,000	14. 540	19. 33	
5. 2,642,000	10. 8,000	15. 27	20. 33	

Page 24 Factors

1. 1, 2, 3, 6	5. 1, 3, 5, 9, 15, 45	9. 1, 13
2. 1, 3, 5, 15	6. 1, 3, 9, 11, 33, 99	10. 1, 2, 7, 14
3. 1, 11	7. 1,2,4,5,10,20,25,50,100	11. 1, 2, 4
4. 1, 3, 9, 27	8. 1, 2, 5, 10	12. 1, 2, 4, 8, 16, 32

Page 25 Greatest Common Factor

1. 10: 1,2,5,10
   15: 1,3,5,15
   GCF: 5

2. 12: 1,2,3,4,6,12
   16: 1,2,4,8,16
   GCF: 4

3. 18: 1,2,3,6,9,18
   36: 1,2,3,4,6,9,12,18,36
   GCF: 18

4. 27: 1,3,9,27
   45: 1,3,5,9,15,45
   GCF: 9

5. 32: 1,2,4,8,16,32
   40: 1,2,4,5,8,10,20,40
   GCF: 8

6. 16: 1,2,4,8,16
   48: 1,2,3,4,6,8,12,16,24,48
   GCF: 16

7. 14: 1,2,7,14
   42: 1,2,3,6,7,14,21,42
   GCF: 14

8. 4: 1,2,4
   26: 1,2,13,26
   GCF: 2

9. 6: 1,2,3,6
   42: 1,2,3,6,7,14,21,42
   GCF: 6

10. 14: 1,2,7,14
    63: 1,3,7,9,21,63
    GCF: 7

11. 9: 1,3,9
    51: 1,3,17,51
    GCF: 3

12. 18: 1,2,3,6,9,18
    45: 1,3,5,9,15,45
    GCF: 9

13. 12: 1,2,3,4,6,12
    20: 1,2,4,5,10,20
    GCF: 4

14. 16: 1,2,4,8,16
    40: 1,2,4,5,8,10,20,40
    GCF: 8

15. 10: 1,2,5,10
    45: 1,3,5,9,15,45
    GCF: 5

16. 18: 1,2,3,6,9,18
    30: 1,3,5,6,10,30
    GCF: 6

Page 26 Prime and Composite Numbers

	Number	Factors	Prime or Composite
1.	24	1, 2, 3, 4, 6, 8, 12, 24	C
2.	33	1, 3, 11, 33	C
3.	25	1, 5, 25	C
4.	42	1, 2, 3, 6, 7, 14, 42	C
5.	14	1, 2, 7, 14	C
6.	35	1, 5, 7, 35	C
7.	47	1, 47	P
8.	49	1, 7, 49	C
9.	18	1, 2, 3, 6, 9, 18	C
10.	56	1, 2, 4, 7, 8, 14, 28, 56	C
11.	71	1, 71	P
12.	52	1, 2, 4, 13, 26, 52	C
13.	19	1, 19	P
14.	31	1, 31	P
15.	26	1, 2, 13, 26	C
16.	16	1, 2, 4, 8, 16	C
17.	81	1, 3, 9, 27, 81	C
18.	44	1, 2, 4, 11, 22, 44	C
19.	90	1, 3, 6, 9, 10, 15, 30, 90	C
20.	45	1, 5, 9, 45	C
21.	13	1, 13	P
22.	9	1, 3, 9	C
23.	12	1, 2, 3, 4, 6, 12	C
24.	27	1, 3, 9, 27	C

Page 28 Prime Factorization

1. 2×5
2. 2×7
3. 5×11
4. $2 \times 5 \times 11$
5. $2 \times 3 \times 3 \times 7$
6. 2×71
7. $2 \times 2 \times 2$
8. 3×7
9. $2 \times 2 \times 2 \times 2 \times 2$
10. $2 \times 2 \times 3 \times 3$
11. 3×17
12. $2 \times 2 \times 3 \times 7$
13. $5 \times 5 \times 5$
14. $2 \times 2 \times 2 \times 2 \times 3$
15. 7×11
16. 5×13
17. $2 \times 2 \times 2 \times 5 \times 5$
18. 7×59
19. $2 \times 2 \times 3$
20. $2 \times 2 \times 2 \times 3$
21. $3 \times 3 \times 5$
22. $2 \times 2 \times 2 \times 3 \times 5$
23. $2 \times 2 \times 13$
24. 7×13
25. $2 \times 3 \times 3$
26. 1×67
27. $2 \times 2 \times 5$
28. 3×5
29. 5×7
30. 2×61

Page 30 Multiples

1. no	4. no	7. yes
2. yes	5. yes	8. no
3. yes	6. no	9. yes

Page 31 Least Common Multiple

1. 30	3. 36	5. 24	7. 28	9. 30	11. 36	13. 90	15. 36	17. 15
2. 48	4. 21	6. 24	8. 18	10. 42	12. 35	14. 24	16. 45	18. 44

Page 32 Whole Number Word Problems

1. 180	3. 115	5. 588	7. 728	9. 24	11. $1,168
2. 6 min	4. 605	6. 92	8. 9,006 ft	10. 70	12. 3,840

Page 33 Determining the Operation

1. $6.25 per friend; Bonus: $7.50	3. $12.00
2. $46.70	4. $4.05

Chapter 1 Review
Pages 34–35

1. four ones	9. 5,922	17. 5	25. C; 1, 5, 25
2. four millions	10. 1,728	18. 8	26. 24
3. four ten thousands	11. 10,396	19. C; 1, 3, 5, 15	27. 45
4. 0	12. 131	20. C; 1, 2, 4, 8, 16	28. 20
5. 6	13. 11	21. P; 1, 19	29. 24
6. 4	14. 1,907	22. C; 1, 2, 4, 5, 10, 20	30. $3
7. 910	15. 3	23. C; 1, 3, 7, 21	31. 98
8. 175	16. 4	24. P; 1, 23	32. 792

Chapter 1 Test
Pages 36–38

1. A	5. D	9. A	13. D	17. A	21. D
2. B	6. A	10. D	14. B	18. C	22. C
3. B	7. B	11. B	15. D	19. C	23. D
4. C	8. C	12. D	16. C	20. C	

Chapter 2 Decimals

Page 40 Place Value: Less Than One

1. 2 hundredths	4. 2 hundredths	7. 2 thousandths	10. 3 thousandths	13. 4 thousandths
2. 2 tenths	5. 2 thousandths	8. 2 thousandths	11. 7 hundredths	14. 6 tenths
3. 2 thousandths	6. 2 thousandths	9. 2 tenths	12. 7 thousandths	15. 6 thousandths

Page 41 Multiplying Decimals by multiples of Ten

1. 27,800	4. 50,700	7. 98.70
2. 1,350	5. 86,500	8. 1,492
3. 187,600	6. 444,000	9. 16,000

Page 42 Multiplying by 0.1 and 0.01

1. 85.1	4. 1.1	7. 17.89
2. 0.65	5. 3.62	8. 45.7
3. 0.08	6. 55.6	9. 0.03

Page 43 Multiplying Decimals by Whole Numbers

1. 53.6	4. 1,346.4	7. 94.8
2. 132.3	5. 772.2	8. 3,591.54
3. 84.4	6. 163.68	9. 1,349

Page 44 Multiplication of Decimals by Decimals

1. 53.20	5. 100.76	9. 217.580	13. 8.20
2. 50.562	6. 31.95	10. 0.6622	14. 2.7927
3. 36.036	7. 0.01716	11. 2.867	15. 3.8610
4. 15.36	8. 28.016	12. 2.041	16. 1.794

Page 45 More Multiplying Decimals by Decimals

1. 0.05535	5. 7.56	9. 32.277	13. 0.1396
2. 0.04124	6. $0.39	10. $1.65	14. 0.2496
3. 1.694	7. 0.045027	11. 0.11828	15. 0.64422
4. 4.5225	8. $0.60	12. 1.1408	16. 0.005056

Page 46 Division of Decimals by Whole Numbers

1. 14.25	3. 3.26	5. 6.25	7. 3.52	9. 18.09	11. 7.25	13. 20.50	15. 17.21
2. 12.36	4. 5.89	6. 4.32	8. 12.24	10. 25.3	12. 2.125	14. 12.6	16. 3.9

Page 47 Division of Decimals by Decimals

1. 52	5. 14.2	9. 85.6	13. 60
2. 879	6. 7300	10. 145.8	14. 50
3. $65.00	7. $167.00	11. 1010	15. 4.25
4. $23.00	8. $20.00	12. 21.25	16. $670.00

Page 48 Ordering Decimals

1. 0.75, 0.705, 0.7, 0.075
2. 0.65, 0.56, 0.5, 0.06
3. 0.95, 0.9, 0.099, 0.09
4. 0.66, 0.6, 0.59, 0.06
5. 0.33, 0.303, 0.3, 0.03
6. 0.5, 0.25, 0.205, 0.02
7. 0.44, 0.4, 0.045, 0.004
8. 0.905, 0.59, 0.509, 0.099
9. 0.111, 0.11, 0.1, 0.01
10. 0.87, 0.8, 0.78, 0.078
11. 0.49, 0.45, 0.41, 0.409
12. 0.754, 0.75, 0.74, 0.7
13. 0.63, 0.07, 0.069, 0.06
14. 0.275, 0.23, 0.208, 0.027
15. 0.05, 0.055, 0.5, 0.59
16. 0.7, 0.72, 0.732, 0.74
17. 0.04, 0.048, 0.408, 0.48
18. 0.09, 0.9, 0.905, 0.95
19. 0.09, 0.1, 0.19, 0.9
20. 0.02, 0.021, 0.2, 0.21

Page 49 Decimal Word Problems

1. $11. 20	3. $9. 99	5. $645. 33	7. 531	9. $896.05
2. $1. 25	4. $7. 55	6. $26. 24	8. 25.38	10. $62.11

Chapter 2 Review
Page 50

1. 4 thousandths	6. 1,655,400	10. 1,987.2	14. 1.4943	18. 0.235
2. 3 hundredths	7. 52.4	11. 1,249.6	15. 0.12587	19. 23
3. 7 tenths	8. 8.56	12. 617.5	16. 320	20. 34.8 pounds
4. 4,506	9. 0.23	13. 0.1145	17. 142	21. 42.3 s
5. 1,958.5				

Chapter 2 Test

Pages 51–52

1. B 5. B 9. A 13. D 17. D

2. A 6. A 10. D 14. C 18. A

3. C 7. B 11. B 15. A 19. A

4. D 8. B 12. C 16. C 20. C

Chapter 3 Fractions

Page 53 Simplifying Fractions

1. $\frac{1}{4}$ 6. $\frac{1}{2}$ 11. $\frac{2}{3}$ 16. $\frac{1}{3}$ 21. $1\frac{1}{6}$ 26. $9\frac{2}{9}$

2. $\frac{4}{5}$ 7. $\frac{7}{11}$ 12. $\frac{1}{4}$ 17. $\frac{1}{3}$ 22. $3\frac{1}{3}$ 27. $15\frac{2}{9}$

3. $\frac{1}{3}$ 8. $\frac{3}{7}$ 13. $\frac{2}{5}$ 18. $\frac{4}{7}$ 23. $4\frac{3}{5}$ 28. $8\frac{2}{7}$

4. $\frac{2}{7}$ 9. $\frac{2}{7}$ 14. $\frac{1}{4}$ 19. $\frac{4}{9}$ 24. $7\frac{1}{2}$ 29. $5\frac{1}{3}$

5. $\frac{1}{7}$ 10. $\frac{3}{13}$ 15. $\frac{1}{2}$ 20. $\frac{2}{3}$ 25. $8\frac{1}{6}$ 30. $10\frac{1}{2}$

Page 54 Simplifying Improper Fractions

1. $2\frac{3}{5}$ 5. $3\frac{1}{6}$ 9. $7\frac{1}{3}$ 13. $1\frac{8}{9}$ 17. $1\frac{3}{4}$ 21. 11

2. $3\frac{2}{3}$ 6. $2\frac{2}{7}$ 10. $3\frac{1}{4}$ 14. $3\frac{3}{8}$ 18. $2\frac{1}{10}$ 22. $3\frac{1}{2}$

3. 4 7. $1\frac{5}{8}$ 11. $7\frac{1}{2}$ 15. $4\frac{4}{7}$ 19. $7\frac{1}{2}$ 23. $6\frac{5}{6}$

4. $1\frac{1}{6}$ 8. $1\frac{4}{5}$ 12. $2\frac{4}{9}$ 16. $1\frac{1}{2}$ 20. $12\frac{2}{5}$ 24. $10\frac{1}{2}$

Page 55 Finding Numerators

1. 6 6. 30 11. 15 16. 4 21. 18 26. 5

2. 18 7. 9 12. 12 17. 9 22. 12 27. 8

3. 8 8. 6 13. 24 18. 12 23. 3 28. 12

4. 21 9. 10 14. 2 19. 14 24. 21 29. 42

5. 45 10. 12 15. 35 20. 22 25. 4 30. 12

Page 56 Adding Fractions

1. $9\frac{2}{9}$ 3. $6\frac{7}{20}$ 5. $11\frac{1}{6}$ 7. $8\frac{1}{12}$ 9. $13\frac{11}{30}$ 11. $6\frac{1}{44}$

2. $5\frac{13}{20}$ 4. $4\frac{1}{8}$ 6. $15\frac{1}{30}$ 8. $13\frac{1}{9}$ 10. $5\frac{11}{14}$ 12. $1\frac{2}{45}$

Page 57 Subtracting Fractions

1. $\frac{5}{36}$
2. $\frac{7}{30}$
3. $\frac{3}{40}$
4. $\frac{7}{20}$
5. $\frac{13}{20}$
6. $\frac{7}{12}$
7. $\frac{2}{15}$
8. $\frac{2}{9}$

Page 58 Subtracting Mixed Numbers from Whole Numbers

1. $8\frac{7}{9}$
2. $1\frac{3}{7}$
3. $12\frac{1}{5}$
4. $\frac{3}{5}$
5. $2\frac{3}{8}$
6. $1\frac{1}{8}$
7. $4\frac{5}{12}$
8. $4\frac{2}{3}$
9. $1\frac{1}{2}$
10. $3\frac{4}{5}$
11. $1\frac{6}{11}$
12. $4\frac{1}{10}$
13. $8\frac{1}{4}$
14. $1\frac{1}{9}$
15. $7\frac{1}{7}$
16. $19\frac{17}{20}$
17. $3\frac{1}{3}$
18. $\frac{2}{5}$
19. $\frac{3}{8}$
20. $4\frac{6}{7}$
21. $7\frac{5}{6}$
22. $\frac{1}{3}$
23. $11\frac{7}{9}$
24. $1\frac{4}{13}$
25. $5\frac{5}{8}$
26. $3\frac{4}{9}$
27. $7\frac{3}{4}$
28. $11\frac{2}{7}$

Page 59 Subtracting Mixed Numbers with Borrowing

1. $2\frac{7}{9}$
2. $\frac{11}{18}$
3. $3\frac{5}{21}$
4. $1\frac{9}{10}$
5. $3\frac{1}{10}$
6. $4\frac{13}{20}$
7. $7\frac{5}{12}$
8. $4\frac{10}{21}$
9. $2\frac{33}{40}$
10. $3\frac{1}{30}$
11. $\frac{17}{36}$
12. $1\frac{11}{30}$
13. $2\frac{23}{30}$
14. $3\frac{7}{8}$
15. $2\frac{7}{9}$
16. $3\frac{1}{2}$
17. $3\frac{1}{2}$
18. $3\frac{5}{6}$
19. $3\frac{23}{28}$
20. $8\frac{11}{20}$

Page 60 Multiplying Fractions with Canceling

1. $\frac{12}{35}$
2. $\frac{3}{20}$
3. $\frac{2}{21}$
4. $\frac{2}{9}$
5. $\frac{4}{45}$
6. $\frac{3}{40}$
7. $\frac{1}{6}$
8. $\frac{1}{4}$
9. $\frac{1}{36}$
10. $\frac{1}{4}$
11. $\frac{1}{3}$
12. $\frac{1}{4}$
13. $\frac{1}{4}$
14. $\frac{1}{14}$
15. $\frac{1}{9}$
16. $\frac{2}{3}$
17. $\frac{3}{4}$
18. $\frac{1}{3}$
19. $\frac{1}{14}$
20. $\frac{1}{3}$
21. $\frac{1}{6}$

Page 61 Changing Mixed Numbers to Improper Fractions

1. $\frac{7}{2}$
2. $\frac{23}{8}$
3. $\frac{29}{3}$
4. $\frac{23}{5}$
5. $\frac{29}{4}$
6. $\frac{69}{8}$
7. $\frac{9}{7}$
8. $\frac{22}{9}$
9. $\frac{31}{5}$
10. $\frac{37}{7}$
11. $\frac{18}{5}$
12. $\frac{75}{8}$
13. $\frac{54}{5}$
14. $\frac{33}{10}$
15. $\frac{29}{7}$
16. $\frac{17}{6}$
17. $\frac{52}{7}$
18. $\frac{61}{9}$
19. $\frac{37}{5}$
20. $\frac{13}{7}$
21. $\frac{4}{1}$
22. $\frac{10}{1}$
23. $\frac{3}{1}$
24. $\frac{2}{1}$
25. $\frac{15}{1}$
26. $\frac{5}{1}$
27. $\frac{6}{1}$
28. $\frac{11}{1}$
29. $\frac{8}{1}$
30. $\frac{16}{1}$

Page 62 Multiplying Mixed Numbers

1. $4\frac{4}{5}$	3. $9\frac{3}{4}$	5. $2\frac{1}{10}$	7. 19	9. 32	11. $1\frac{2}{3}$	13. $14\frac{1}{2}$	15. $6\frac{1}{2}$
2. $1\frac{2}{7}$	4. $17\frac{1}{2}$	6. $2\frac{6}{7}$	8. $9\frac{1}{3}$	10. $8\frac{1}{4}$	12. $1\frac{3}{4}$	14. $5\frac{3}{4}$	16. 5

Page 63 Dividing Fractions

1. $1\frac{1}{2}$	3. $\frac{13}{18}$	5. $1\frac{3}{11}$	7. $12\frac{1}{2}$	9. $\frac{1}{5}$	11. $2\frac{10}{13}$	13. $4\frac{2}{3}$	15. $\frac{7}{8}$
2. $3\frac{1}{3}$	4. 4	6. $3\frac{1}{7}$	8. $4\frac{1}{6}$	10. $2\frac{1}{2}$	12. $2\frac{2}{9}$	14. $\frac{5}{9}$	16. 7

Page 64 Estimating Multiplying Fractions

1. 300 2. 100 3. 300 4. 100 5. 300 6. 700 7. 700 8. 800 9. 500

Page 64 Estimating Reducing Fractions

1. $\frac{2}{5}$ 2. $\frac{1}{8}$ 3. $\frac{1}{7}$ 4. $\frac{1}{9}$ 5. $\frac{1}{10}$ 6. 5 7. $\frac{1}{4}$ 8. $\frac{1}{4}$ 9. $\frac{1}{90}$

Page 65 Comparing Fractions

1. $<$	3. $<$	5. $>$	7. $>$	9. $<$	11. $>$
2. $>$	4. $<$	6. $>$	8. $>$	10. $<$	12. $<$

Page 66 Changing Fractions to Decimals

1. 0.8	5. 0.1	9. 0.6	13. $0.\overline{77}$	17. 0.1875
2. $0.\overline{6}$	6. 0.625	10. 0.7	14. 0.9	18. 0.75
3. 0.5	7. $0.8\overline{3}$	11. $0.\overline{36}$	15. 0.25	19. $0.\overline{8}$
4. $0.\overline{55}$	8. $0.1\overline{6}$	12. $0.\overline{11}$	16. 0.375	20. $0.41\overline{6}$

Page 67 Changing Mixed Numbers to Decimals

1. $5.\overline{6}$	5. $30.\overline{3}$	9. 6.8	13. 7.25	17. 10.1
2. $8.\overline{45}$	6. 3.5	10. 13.5	14. $12.\overline{3}$	18. 20.4
3. 15.6	7. 1.875	11. 12.8	15. 1.625	19. 4.9
4. $13.\overline{6}$	8. 4.09	12. 11.625	16. 2.75	20. $5.\overline{36}$

Page 67 Changing Decimals to Fractions

1. $\frac{11}{20}$
2. $\frac{3}{5}$
3. $\frac{3}{25}$
4. $\frac{9}{10}$
5. $\frac{3}{4}$
6. $\frac{41}{50}$
7. $\frac{3}{10}$
8. $\frac{21}{50}$
9. $\frac{71}{100}$
10. $\frac{21}{50}$
11. $\frac{14}{25}$
12. $\frac{6}{25}$
13. $\frac{7}{20}$
14. $\frac{24}{25}$
15. $\frac{1}{8}$
16. $\frac{3}{8}$

Page 68 Changing Decimals with Whole Numbers to Mixed Numbers

1. $7\frac{1}{8}$
2. $99\frac{1}{2}$
3. $2\frac{13}{100}$
4. $5\frac{1}{10}$
5. $16\frac{19}{20}$
6. $3\frac{5}{8}$
7. $4\frac{21}{50}$
8. $15\frac{21}{25}$
9. $6\frac{7}{10}$
10. $45\frac{17}{40}$
11. $15\frac{4}{5}$
12. $8\frac{4}{25}$
13. $13\frac{9}{10}$
14. $32\frac{13}{20}$
15. $17\frac{1}{4}$
16. $9\frac{41}{50}$

Page 68 Fraction Word Problems

1. $2\frac{53}{60}$
2. $4\frac{1}{2}$
3. 96
4. 750
5. $13\frac{1}{3}$
6. $1\frac{5}{6}$

Chapter 3 Review
Pages 69–70

1. $2\frac{1}{2}$
2. $4\frac{4}{5}$
3. $1\frac{1}{3}$
4. $4\frac{2}{3}$
5. $\frac{1}{3}$
6. $\frac{1}{4}$
7. $\frac{2}{3}$
8. $\frac{2}{3}$
9. $\frac{51}{10}$
10. $\frac{7}{1}$
11. $\frac{18}{5}$
12. $\frac{20}{3}$
13. $1\frac{1}{3}$
14. $10\frac{7}{8}$
15. $4\frac{7}{15}$
16. $\frac{4}{7}$
17. $4\frac{7}{8}$
18. $2\frac{7}{12}$
19. $7\frac{3}{8}$
20. $4\frac{9}{10}$
21. $4\frac{2}{3}$
22. $4\frac{3}{4}$
23. 8
24. $\frac{5}{9}$
25. $\frac{5}{8}$
26. $2\frac{4}{7}$
27. $\frac{1}{3}$
28. $\frac{5}{12}$
29. 40
30. $\frac{1}{2}$
31. 20
32. 18
33. 10
34. 56
35. 27
36. 20
37. 24
38. 33
39. $\frac{11}{20}$
40. $\frac{21}{25}$
41. $\frac{8}{25}$
42. $7\frac{3}{8}$
43. $9\frac{3}{5}$
44. $13\frac{1}{4}$
45. 5.12
46. 0.07
47. 10.67
48. $>$
49. $<$
50. $<$
51. $<$
52. $12\frac{5}{6}$ miles
53. $2\frac{17}{60}$ miles
54. $17\frac{1}{2}$ gallons

Chapter 3 Test
Pages 71–72

1. C
2. B
3. C
4. A
5. D
6. B
7. C
8. C
9. C
10. C
11. A
12. C
13. D
14. C
15. C
16. B

Chapter 4 Percents

Page 73 Changing Percents to Decimals and Decimals to Percents

1. 0.18
2. 0.23
3. 0.09
4. 0.63
5. 0.04
6. 0.45
7. 0.02
8. 1.19
9. 0.07
10. 0.55
11. 0.80
12. 0.17
13. 0.66
14. 0.13
15. 0.05
16. 0.25
17. 4.10
18. 0.01
19. 0.50
20. 0.99
21. 1.07
22. 15%
23. 62%
24. 153%
25. 22%
26. 35%
27. 37.5%
28. 64.8%
29. 4.4%
30. 58%
31. 86%
32. 29%
33. 6%
34. 48%
35. 308.9%
36. 4.2%
37. 37.5%
38. 509%
39. 75%
40. 30%
41. 290%
42. 6%

Page 74 Changing Percents to Fractions

1. $\frac{1}{2}$
2. $\frac{13}{100}$
3. $\frac{11}{50}$
4. $\frac{19}{20}$
5. $\frac{13}{25}$
6. $\frac{63}{100}$
7. $\frac{3}{4}$
8. $\frac{91}{100}$
9. $\frac{9}{50}$
10. $\frac{3}{100}$
11. $\frac{1}{4}$
12. $\frac{1}{20}$
13. $\frac{4}{25}$
14. $\frac{1}{100}$
15. $\frac{79}{100}$
16. $\frac{2}{5}$
17. $\frac{99}{100}$
18. $\frac{3}{10}$
19. $\frac{3}{20}$
20. $\frac{21}{25}$

Page 74 Changing Fractions to Percents

1. 20%
2. 62.5%
3. 43.75%
4. 37.5%
5. 18.75%
6. 19%
7. 10%
8. 80%
9. 93.75%
10. 75%
11. 12.5%
12. 31.25%
13. 6.25%
14. 25%
15. 4%
16. 75%
17. 40%
18. 64%

Page 75 Changing Percents to Mixed Numbers

1. $1\frac{1}{2}$
2. $1\frac{13}{100}$
3. $2\frac{11}{50}$
4. $3\frac{19}{20}$
5. $2\frac{13}{25}$
6. $1\frac{63}{100}$
7. $2\frac{3}{4}$
8. $1\frac{91}{100}$
9. $1\frac{2}{25}$
10. $4\frac{53}{100}$
11. $2\frac{1}{20}$
12. $4\frac{1}{20}$
13. $5\frac{4}{25}$
14. $1\frac{61}{100}$
15. $1\frac{79}{100}$
16. $3\frac{2}{5}$
17. $1\frac{99}{100}$
18. 3
19. $1\frac{1}{4}$
20. $3\frac{21}{25}$

Page 75 Changing Mixed Numbers to Percents

1. 550%
2. 875%
3. 100%
4. 325%
5. 487.5%
6. 300%
7. 130%
8. 620%
9. 400%
10. 252%
11. 112.5%
12. 200%
13. 118.75%
14. 106.25%
15. 500%
16. 480%
17. 340%
18. 600%

Page 76 Comparing the Relative Magnitude of Numbers

1. <
2. =
3. =
4. >
5. <
6. >
7. >
8. >
9. <
10. =
11. <
12. =
13. =
14. =
15. >
16. <
17. <
18. >
19. <
20. >
21. <
22. =
23. >
24. <
25. =

Page 78 Modeling Percentages

1. 10%
2. 35%
3. 68%
4. 3%
5. 50%
6. 9%
7. 22%
8. 37%
9. 75%
10. 33.33%
11. 66.67%

Page 79 Representing Rational Numbers Graphically

1.	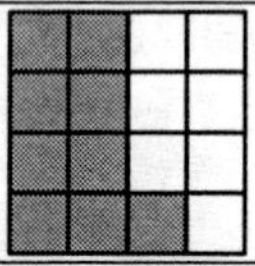	$\frac{9}{16}$	0.5625	56.25%	5.	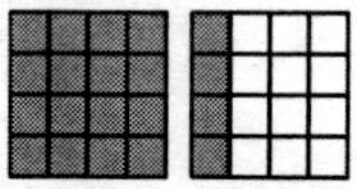	$1\frac{1}{4}$	1.25	125%
2.	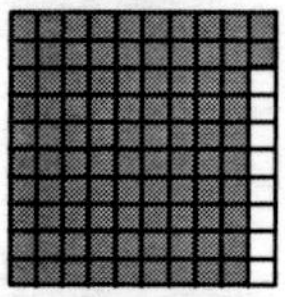	$\frac{23}{25}$	0.92	92%	6.	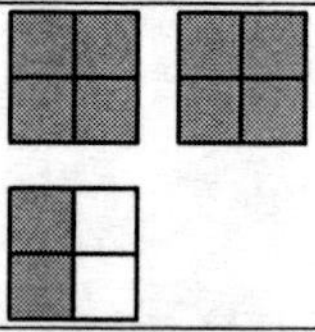	$2\frac{1}{2}$	2.5	250%
3.	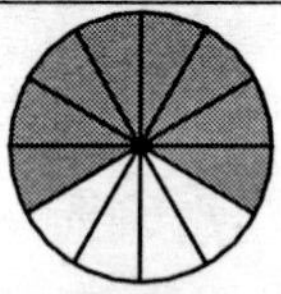	$\frac{2}{3}$	$0.\overline{6}$	$66.\overline{6}\%$	7.	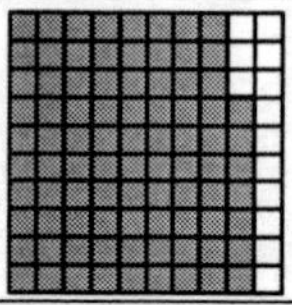	$\frac{87}{100}$	0.87	87%
4.	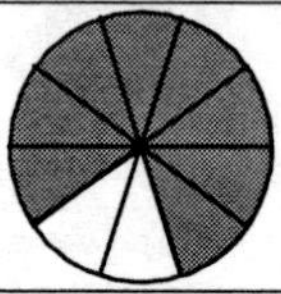	$\frac{4}{5}$	0.8	80%	8.	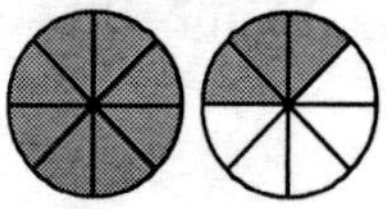	$1\frac{3}{8}$	1.375	137.5%

Page 80 Percent Word Problems

1. 75%
2. 20%
3. 30%
4. 80%
5. 25%
6. 16%
7. 48%
8. 75%
9. 72%
10. 92%
11. 40%
12. 85%
13. 87.5%
14. 52%
15. 82.4%
16. 95%
17. 68%
18. 15%

Page 81 Missing Information

1. What was the total number of candies Amin had?
2. What is the length of Patrick's yard?
3. How many hours did Yoko work?
4. How many ounces is the bath bar?
5. What is the sales tax rate?
6. How many people are in the Portes family?
7. When did the kudzu start growing?
8. What time did Bethany leave for her sister's house?
9. How many hours did Terrance work?
10. What is the amount of the employee discount?
11. What was the price per gallon of the gas?
12. How many work days are in the month?

Page 82 Sales Tax

1. \$44.94
2. \$18544.70
3. \$6.36
4. \$12.60
5. \$37.86
6. \$1.87
7. \$116.38
8. \$19.08
9. \$2.46
10. \$97.15

Chapter 4 Review
Page 83

1. 0.45
2. 2.19
3. 0.22
4. 0.0125
5. 52%
6. 64%
7. 109%
8. 62.5%
9. 165%
10. 565%
11. $\frac{1}{4}$
12. $\frac{3}{100}$
13. $\frac{17}{25}$
14. $1\frac{1}{50}$
15. 90%
16. 31.25%
17. 12.5%
18. 25%
19. 100%
20. 77%
21. 40%
22. 37.5%

Chapter 4 Test
Pages 84–85

1. A
2. D
3. B
4. C
5. C
6. A
7. C
8. B
9. B
10. B
11. B
12. D
13. B
14. D
15. D

Chapter 5 Introduction to Algebra

Page 87 Order of Operations

1. 20
2. 12
3. 7
4. 1
5. 35
6. 4
7. 3
8. 23
9. 8
10. 48
11. 10
12. 33
13. 15
14. 83
15. 8
16. 2
17. 1
18. 7
19. 7
20. 25

Page 88 Substituting Numbers for Variables

1. 10
2. 11
3. 3
4. 16
5. 8
6. 10
7. 21
8. 1
9. 41
10. 0
11. 63
12. 14
13. 26
14. 20
15. 2
16. 2
17. 2
18. 80
19. 21
20. 60
21. 3
22. 38
23. 35
24. 21

Page 90 Understanding Algebra Word Problems

1. C
2. D
3. A
4. B
5. D
6. C
7. E
8. B
9. A
10. B
11. E
12. C
13. $x - 3$
14. $\frac{y}{10}$
15. $t + 5$
16. $n - 14$
17. $5k$
18. $z + 12$
19. $2b$
20. $x + 1$
21. $\frac{t}{4}$
22. $\frac{y}{2}$
23. $2b$
24. $3y$
25. $n + 4$
26. $t - 6$
27. $\frac{18}{x}$
28. $10 - n$
29. $3 + p$
30. $4m$
31. $y - 20$
32. $5x$

Chapter 5 Review
Page 91

1. 10
2. 3
3. 1
4. 5
5. 6
6. 1
7. 14
8. 7
9. 6
10. 2
11. 17
12. 3
13. 4
14. 7
15. 10
16. 9
17. 1
18. 1

Chapter 86 Test
Page 92

1. B
2. A
3. A
4. D
5. C
6. C
7. A
8. C

Chapter 6 Measurement

Page 94 Approximate English Measure

1. C
2. F
3. E
4. G
5. E
6. G
7. D
8. B
9. H
10. A

Page 95 Converting Units of Measure

1. 4 lb 4 oz
2. 3 cups 4 oz
3. 4 wk 3 days 6 hr
4. 2 pt 1 cup
5. 3 hr 25 min 2 sec
6. 2 gal 3 qt 1 pt
7. 6 yd 2 ft 6 in
8. 3 wk 2 days 12 hrs
9. 3 ft 6 in
10. 3 lb 1 oz
11. 1 day 3 min 34 sec
12. 5 days 7 hr 15 min

Page 96 Converting in the Customary System

1. 2
2. 1
3. 9
4. 8
5. 10
6. 4
7. 16
8. 72
9. 5
10. $\frac{1}{2}$
11. 32
12. 4
13. 48
14. 1.5
15. 24
16. 2.5
17. 1.25
18. $1\frac{1}{3}$
19. 1
20. 1.5
21. 16
22. 36
23. 84
24. 0.875
25. 0.625
26. 2
27. 4
28. 32
29. 3
30. 24

Page 98 Estimating Metric Measurements

1. B
2. A
3. A
4. D
5. C
6. A
7. D
8. C
9. D
10. B
11. A
12. C
13. B
14. C

Page 99 Converting Units within the Metric System

1. 0.035 g
2. 6,000 m
3. 0.0215 L
4. 0.49 cm
5. 5,350,000 mL
6. 0.0000321 kg
7. 0.1564 km
8. 2.5 cg
9. 17,500 mL
10. 0.0042 kg
11. 6 dL
12. 41,700 cg
13. 0.182 L
14. 812 cm
15. 0.723 mm
16. 3 L
17. 5,060 mg
18. 0.1058 cL
19. 4.3 km
20. 205.7 cm
21. 0.5643 kg

Chapter 6 Review
Page 100

1. pound
2. inches
3. liters
4. milligrams
5. 32
6. 4,200
7. 126
8. 6.8
9. $2\frac{1}{4}$
10. 0.00073
11. 0.12 km
12. 9,000 mg
13. 20 L
14. 0.0015 g
15. 150 mm
16. 5,000 mL
17. 5 g
18. 0.055 L
19. 0.3 m

Chapter 6 Test
Page 101

1. A
2. D
3. B
4. C
5. C
6. A
7. B
8. C

Chapter 7 Plane Geometry

Page103 Estimating Area

1. 48 units^2
2. 16 units^2
3. 19 units^2
4. 21 units^2

Page 104 Area of Squares and Rectangles

1. 100 ft^2
2. 10 cm^2
3. 36 in^2
4. 180 in^2
5. 36 ft^2
6. 50 cm^2
7. 8 ft^2
8. 40 in^2
9. 144 ft^2
10. 84 cm^2
11. 8 ft^2
12. 42 cm^2

Page 106 Area of Parallelograms

1. 132 in^2
2. 48 in^2
3. 154 in^2
4. 60 cm^2

Page 108 Area of Triangles

1. 6 in^2
2. 36 cm^2
3. 21 ft^2
4. 72 cm^2
5. 3 ft^2
6. 160 cm^2
7. 52.5 m^2
8. 31.5 in^2
9. 2 ft^2
10. 12 ft^2
11. 75 ft^2
12. 15 m^2

Page 109 Circumference

1. 50.24 in
2. 43.96 ft
3. 6.28 cm
4. 37.68 m
5. 25.12 ft
6. 18.84 ft
7. 37.68 in
8. 18.84 m
9. 31.4 cm
10. 50.24 in

Page 110 Comparing Diameter, Radius, and Circumference in a Circle

1. 4 cm
2. 8 cm
3. circumference and diameter
4. 25.12
5. π
6. infinitely
7. yes
8. $\frac{22}{7}$

Page 111 Area of a Circle

1. 78.5 in^2
2. 200.96 ft^2
3. 50.24 cm^2
4. 28.26 m^2
5. 18 ft, 254.34 ft^2
6. 2 in, 12.56 in^2
7. 16 cm, 200.96 cm^2
8. 10 ft, 314 ft^2
9. 28 m, 615.44 m^2
10. 9 cm, 254.34 cm^2
11. 24 ft, 452.16 ft^2
12. 3 in, 28.26 in^2

Page 113 Two-Step Area Problems

1. 525 ft^2
2. 112 in^2
3. 452 cm^2
4. 73 ft^2
5. 422 in^2
6. 12.5 m^2
7. 2,500 cm^2
8. 216 m^2

Page 114 Similar and Congruent

1. N
2. S
3. C
4. C
5. N
6. C
7. S
8. C
9. C
10. N
11. S
12. N
13. C
14. N

Page 116 Congruent Figures

1. congruent, all corresponding angles and sides are congruent
2. not congruent, corresponding angles are not congruent
3. congruent, all corresponding angles and sides are congruent
4. not congruent, corresponding angles are not congruent
5. not congruent, corresponding sides are not equal
6. not congruent, corresponding sides are not congruent

Chapter 7 Review
Page 117

1. 25 $units^2$
2. 32 $units^2$
3. $C = 6.28$ ft, $A = 3.14$ ft^2
4. 170 in^2
5. 66 cm^2
6. 24 in^2
7. 12 ft^2

Chapter 7 Test
Pages 118–119

1. B
2. D
3. A
4. D
5. A
6. B
7. C
8. C
9. B
10. D
11. C
12. C

Chapter 8 Solid Geometry

Page 121 Understanding Volume

1. 8 ft^3
2. 21 cm^3
3. 60 in^3
4. 288 in^3
5. 200 mm^3
6. 360 ft^3

Page 122 Volume of Rectangular Prisms

1. 72 ft^3
2. 1,200 m^3
3. 240 cm^3
4. 1,500 m^3
5. 90 ft^3
6. 4,480 in^3
7. 675 in^3
8. 336 cm^3
9. 18 m^3

Page 123 Volume of Cubes

1. 27 cm^3
2. 216 cm^3
3. 729 cm^3
4. 64
5. 8 cm^3
6. 4 in
7. 343 in^3
8. 64 ft^3
9. 1,728 in^3

Chapter 8 Review
Page 124

1. $V = 18$ cm^3
2. 36,000 in^3
3. 64 in^3
4. 896 ft^3
5. 560 m^3
6. 72 in^3
7. 4 in^3

Chapter 8 Test
Page 125

1. C 2. C 3. C 4. A 5. B 6. B 7. D 8. A 9. C 10. B

Chapter 9 Data Interpretation

Page 128 Reading Tables

1. 12 2. 36 3. 37 4. 16 5. 15 6. 28

Page 130 Bar Graphs

1. 155
2. 28
3. 285
4. 75
5. 103
6. 30
7. 17
8. C
9. 90
10. 52

Page 132 Line Graphs

1. B 2. D 3. D 4. A

Page 134 Circle Graphs

1. \$16
2. \$20
3. \$40
4. \$4
5. 300
6. 150
7. 280
8. 50
9. 150
10. 70

Page 135 Comparing Types of Graphs

1. circle graph	3. line graph	5. circle graph	7. Jim
2. line graph	4. circle graph	6. yes	

Chapter 9 Review
Pages 136–137

1. 33	5. 37.6 million metric tons	9. Africa
2. 2	6. 14.8 million metric tons	10. 80 million metric tons
3. 40	7. 1.8 billion	11. 1983
4. 11.2 million metric tons	8. 144 million	12. 20 million metric tons

Chapter 9 Test
Pages 138–139

1. C	2. B	3. C	4. C	5. A	6. C	7. D	8. C

Practice Test 1

Pages 141–149

1. C	8. C	15. A	22. C	29. B	36. B	43. B	50. D	57. B	64. A
2. C	9. B	16. B	23. B	30. C	37. A	44. C	51. C	58. B	65. C
3. B	10. D	17. A	24. B	31. B	38. A	45. B	52. B	59. A	66. D
4. A	11. A	18. B	25. A	32. C	39. B	46. A	53. D	60. A	67. B
5. C	12. A	19. B	26. A	33. C	40. C	47. C	54. C	61. C	68. B
6. D	13. D	20. C	27. A	34. C	41. B	48. C	55. A	62. C	69. D
7. D	14. B	21. D	28. D	35. C	42. B	49. D	56. B	63. A	70. B

Practice Test 2

Pages 150–161

1. B	8. A	15. D	22. A	29. A	36. D	43. A	50. A	57. D	64. A
2. D	9. C	16. D	23. B	30. C	37. B	44. D	51. D	58. C	65. C
3. C	10. D	17. C	24. C	31. B	38. A	45. B	52. C	59. B	66. C
4. D	11. A	18. A	25. C	32. D	39. C	46. A	53. A	60. C	67. B
5. A	12. C	19. D	26. D	33. C	40. B	47. C	54. C	61. A	68. C
6. C	13. B	20. B	27. B	34. B	41. D	48. D	55. D	62. B	69. A
7. A	14. B	21. D	28. A	35. A	42. C	49. B	56. B	63. A	70. A

Product Order Form

Please fill this form out completely and fax it to 1-866-827-3240

Purchase Order #: ____________ Date: ____________

Contact Person: ____________

School Name (and District, if any): ____________

Billing Address: ____________ Street Address: ☐ same as billing ____________

Attn: ____________ Attn: ____________

Phone: ____________ E-Mail: ____________

Credit Card #: ____________ Exp Date: ____________

Authorized Signature: ____________

Order Number	Product Title	Pricing* 5 books	Qty	Pricing 30+ books	Qty	Total Cost
GA3-M0607	Mastering the Georgia 3rd Grade CRCT in Math	$59.95 (1 set of 5 books)		$269.70 (1 set of 30 books)		
GA3-R0607	Mastering the Georgia 3rd Grade CRCT in Reading	$59.95 (1 set of 5 books)		$269.70 (1 set of 30 books)		
GA5-M0806	Mastering the Georgia 5th Grade CRCT in Math	$59.95 (1 set of 5 books)		$269.70 (1 set of 30 books)		
GA5-R1206	Mastering the Georgia 5th Grade CRCT in Reading	$59.95 (1 set of 5 books)		$269.70 (1 set of 30 books)		
GA5-S1107	Mastering the Georgia 5th Grade CRCT in Science	$59.95 (1 set of 5 books)		$269.70 (1 set of 30 books)		
GA6-M0305	Mastering the Georgia 6th Grade CRCT in Math	$59.95 (1 set of 5 books)		$269.70 (1 set of 30 books)		
GA6-R0108	Mastering the Georgia 6th Grade CRCT in Reading	$59.95 (1 set of 5 books)		$269.70 (1 set of 30 books)		
GA6-S1206	Mastering the Georgia 6th Grade CRCT in Science	$59.95 (1 set of 5 books)		$269.70 (1 set of 30 books)		
GA6-H0208	Mastering the Georgia 6th Grade CRCT in Social Studies	$59.95 (1 set of 5 books)		$269.70 (1 set of 30 books)		
GA7-M0305	Mastering the Georgia 7th Grade CRCT in Math	$59.95 (1 set of 5 books)		$269.70 (1 set of 30 books)		
GA7-R0707	Mastering the Georgia 7th Grade CRCT in Reading	$59.95 (1 set of 5 books)		$269.70 (1 set of 30 books)		
GA7-S1206	Mastering the Georgia 7th Grade CRCT in Science	$59.95 (1 set of 5 books)		$269.70 (1 set of 30 books)		
GA7-H0208	Mastering the Georgia 7th Grade CRCT in Social Studies	$59.95 (1 set of 5 books)		$269.70 (1 set of 30 books)		
GA8-M0305	Passing the Georgia 8th Grade CRCT in Math	$59.95 (1 set of 5 books)		$269.70 (1 set of 30 books)		
GA8-L0505	Passing the Georgia 8th Grade CRCT in Language Arts	$59.95 (1 set of 5 books)		$269.70 (1 set of 30 books)		
GA8-R0505	Passing the Georgia 8th Grade CRCT in Reading	$59.95 (1 set of 5 books)		$269.70 (1 set of 30 books)		
GA8-S0707	Mastering the Georgia 8th Grade CRCT in Science	$59.95 (1 set of 5 books)		$269.70 (1 set of 30 books)		
GA8-H0607	Mastering the Georgia 8th Grade CRCT in GA Studies	$59.95 (1 set of 5 books)		$269.70 (1 set of 30 books)		
GA8-W0907	Passing the Georgia 8th Grade Writing Assessment	$59.95 (1 set of 5 books)		$269.70 (1 set of 30 books)		

1-7-08 *Minimum order is 1 set of 5 books of the same subject.

Subtotal	
Shipping & Handling 12%	
Total	

American Book Company ● PO Box 2638 ● Woodstock, GA 30188-1383
Toll-Free Phone: 1-888-264-5877 ● Toll-Free Fax: 1-866-827-3240 ● Web Site: www.americanbookcompany.com